Telecommunications and Geography

For Tovy, Miri, and Noga

TELECOMMUNICATIONS AND GEOGRAPHY

AHARON KELLERMAN

Belhaven Press
London and New York

Copublished in the Americas by Halsted Press, an imprint of
John Wiley & Sons, Inc., New York

Belhaven Press
(a division of Pinter Publishers)
25 Floral Street, Covent Garden, London, WC2E 9DS, United Kingdom

First published in 1993

© Aharon Kellerman

Apart from any fair dealing for the purposes of research or private study, or criticism or review, as permitted under the Copyright, Designs and Patents Act, 1988, this publication may not be reproduced, stored or transmitted in any form or by any means or process without the prior permission in writing of the copyright holders or their agents. Except for reproduction in accordance with the terms of licences issued by the Copyright Licensing Agency, photocopying of whole or part of this publication without the prior written permission of the copyright holders or their agents in single or multiple copies whether for gain or not is illegal and expressly forbidden. Please direct all enquiries concerning copyright to the Publishers at the address above.

Co-published in the Americas by Halsted Press, an imprint of
John Wiley & Sons, Inc., 605 Third Avenue, New York, NY 10158–0012

Aharon Kellerman is hereby identified as the author of this work as provided under Section 77 of the Copyright, Designs and Patents Act, 1988.

British Library Cataloguing in Publication Data
A CIP catalogue record for this book is available from the British Library
ISBN 1 85293 247 3

Library of Congress Cataloging-in-Publication Data
A CIP catalog record for this book is available from the Library of Congress

ISBN 0–470–22034–1 (in the Americas only)

Typeset by Mayhew Typesetting, Rhayader, Powys
Printed and bound in Great Britain by Biddles Ltd of Guildford and King's Lynn

Politically, the conveying of information is the most important of all communication performances.

(Friedrich Ratzel, *Politische Geografie*, 1897, p.411).

Well, without [the] 'communication' there can be no 'geography of'. You cannot have a *geography* of anything that is unconnected. No connections, no geography.

(Peter Gould, 'Dynamic structures of geographic space', 1991, p.4).

Contents

List of figures xi
List of tables xiii
Preface xv

PART 1: GEOGRAPHICAL CONCEPTS FOR TELECOMMUNICATIONS

Chapter 1
Introduction: Telecommunications, Information, and Geography 3
  Telecommunications: meanings and significance 3
    Telecommunications and computers 4
  Telecommunications, services, and information economies/societies 6
    Telecommunications and service economies 6
    Telecommunications and information economies 9
  The geography of telecommunications 12
    Early developments in the geography of telecommunications 13
    Basic notions for the geography of telecommunications 15
  Conclusion 18

Chapter 2
Telecommunications as Material Entity and Spatial Artefact 20
  Transmission media 20
    The development of telecommunications 21
    Recent trends in the telecommunications industry 26

Sociospatial characteristics and impacts of
telecommunications   30
Networks   35
    Cable networks   36
    Microwave and satellite communications   38
    Maritime cables   47
Nodes   53
    Smart buildings   54
    Teleports   55
    Telecottages   57
Conclusion   58

Chapter 3
Spatial Dynamics of Telecommunications   60
Flows   60
    Information types   61
    Flow characteristics   64
    Flow barriers   68
    Flow-induced patterns   72
The diffusion and penetration of telecommunications
innovations   73
    The diffusion process   74
    Comparative diffusion trends   75
    The telephone   76
    The fax (facsimile)   77
    Home computers   80
    Cable television   81
Telecommunications versus transportation:
substitutability and complementarity   83
    Telecommuting   83
    Teleshopping   84
    Business calls   85
    Teleconferencing   86
Conclusion: The telecommunications cycle   86

**PART 2: TELECOMMUNICATIONS IN
GEOGRAPHICAL CONTEXTS**

Chapter 4
Telecommunications and Cities   93
Spatial concentration or dispersion?   94
    The antipolitan approach   95

Balanced views on the wired city	96
The global urban system	97
General concepts	98
The global urban hierarchy	98
The development of the system	106
The location of producer services	111
Office suburbanization	111
CBD specialization	113
Transfer to other cities	114
Conclusion	115

Chapter 5
Telecommunications and Regional Development   116

The role and context of telecommunications in regional development	116
The role of telecommunications	117
Telecommunications and interregional inequality	118
Economic sectors	120
Manufacturing	121
Services	121
Patterns and problems in selected regions	123
Less favoured regions (LFRs) in Europe	124
Rural America	125
Less developed countries (LDCs)	128
Conclusion	129

Chapter 6
National Differences in Telecommunications   130

The global distribution of telephones	130
The big gap	130
Teledensities	131
National economic development and telecommunications	134
The correlation between measures of development and teledensities	135
The role of telecommunications in national economic development	138
From 'natural monopoly' to full competition	139
Organizational and ownership patterns for telecommunications services	140
The emergence of organizational and ownership patterns	144
Growth implications of the organizational transitions in telecommunications services	148

x Contents

    Geographical implications of changing ownership
      patterns   150
    Conclusion   151

Chapter 7
International Telecommunications   **153**
    Basic dimensions of international telecommunications   155
      The importance of international telecommunications   155
      Movement aspects of international telecommunications   157
    The regulation of international telecommunications   162
      The International Telecommunication Union (ITU)   163
      International accounting for telecommunications   165
    Public traffic: patterns and factors   167
      The dominance of world cores   169
      Cultural patterns   174
      Colonial imprints   176
      International telecommunications and other international
      movements   177
    Dedicated international networks   180
      Transborder data flows (TBDF)   180
      Transnational corporations (TNCs) and banks   181
      Offshore back offices   182
      Electronic mail networks   183
    Conclusion   185

Chapter 8
Conclusion   **187**
    Telecommunications, concentration and dispersion   187
    American leadership in telecommunications   190
    Directions for a geographical theory of telecommunications   192
    Telecommunications and geography: future scenarios   195

References   198
Glossary   216
Index   218

List of figures

1.1	Telecommunications in information and service contexts.	6
1.2	The evolution of service economies.	8
1.3	Percentage contribution of economic sectors to US national income 1950–80.	11
1.4	Telephone convergence between New York and San Francisco: (a) time-space convergence; (b) cost-space convergence.	16
1.5	Approaches to the geography of telecommunications.	18
2.1	Sociospatial impacts of telecommunications.	31
2.2	Basic topologies for spatial networks of telephone systems.	37
2.3	Major telephone links in France, 1984.	39
2.4	The allocation of the geostationary orbit for C-band satellites, 1992.	42
2.5	The allocation of the geostationary orbit for Ku-band satellites, 1992.	44
2.6	Existing and planned maritime cables, 1988.	49
2.7	Fibre-optics maritime cables, 1992.	50
2.8	Estimated capacity of transoceanic cable and satellite systems 1986–96.	52
2.9	Major teleports, 1986.	56
2.10	Telecottages in Scandinavia, 1988.	57
3.1	Dimensions of information flows through telecommunications.	62
3.2	The ratio of toll to local calls in the US (toll=1), 1900–87.	65

List of figures

3.3	The average lengths of incoming and outgoing US international phone calls, 1964–89.	67
3.4	US Local Access and Transport Areas (LATAs), 1984.	71
3.5	Growth of the AT&T toll system, 1890–1904.	78
3.6	Sales of facsimile machines and telexes in Europe, 1984–92.	79
3.7	Annual sales of personal computers in the US, 1976–91.	81
3.8	Subscribers for cable television in the US, 1964–90.	82
3.9	The telecommunications cycle.	87
4.1	Telecommunications and business services in an urban context.	107
4.2	Major centres of foreign bank offices in the US, 1986.	110
5.1	Telecommunications and regional development.	118
6.1	European teledensities by the number of main lines per 100 inhabitants, 1990.	134
6.2	Regression between per capita GNP and teledensities for selected countries, 1990.	137
6.3	Waiting applications for telephone lines and new installations in Israel, 1984–91.	149
6.4	International comparisons of telecommunications traffic and real revenue growth, 1983–88.	150
7.1	Dimensions of international telecommunications.	154
7.2	The role of international telecommunications in the attempted coup in the Soviet Union, August 1991.	156
7.3	World zones for country first dial digit.	164
7.4	The shares of the five leading industrial nations in US outgoing international telephones calls, 1961–90.	173
7.5	The shares of four world regions in US outgoing international telephone calls to countries ranked 6–15, 1961– 90.	174
7.6	The EARN network (25 April 1988).	185

List of tables

1.1	Percentage information occupations in the labour force of selected OECD countries 1951–81.	11
2.1	Major innovations and developments in telecommunications 1837–1988.	22
2.2	Time-lags between invention and implementation of telecommunications innovations.	23
2.3	Time-lags between innovation of telecommunications devices and technology integration.	23
2.4	Phases of telecommunications infrastructure.	24
2.5	Communications satellites, 1990.	46
3.1	US household penetration of major telecommunications technologies 1990.	76
4.1	The shares of New York, London, and Tokyo in global capital markets (in %).	102
4.2	The shares of leading capital markets in the movements of people, commodities, and information (in %).	103
5.1	The role of telecommunications in regional development.	117
5.2	The percentage distribution of population and telephones in urban and rural areas.	128
6.1	Teledensities by the number of main lines per 100 inhabitants 1990.	132
6.2	Various ownership options for telecommunications services.	142
7.1	Types and characteristics of international movements.	158

List of tables

7.2	The three most frequently called countries for selected nations 1977–87.	170
7.3	The ranks and shares of G7 countries in phone calls made from G7 countries in 1986.	171
7.4	The shares of the 15 most frequently called countries from the US 1990 (in % by number of calls).	175
8.1	Major geographical impacts of telecommunications.	193

Preface

'Telecommunications' sounds new, but its initial invention is old. The telegraph was introduced in the year 1837, followed by the invention of the telephone in 1876. Still though, contemporary and more sophisticated telecommunications constitutes one of the main cornerstones of futurism, claiming a predominance of information exchanges in both households and economic activities (e.g. Toffler, 1981). Reality tends to present more modest trends in various aspects in this regard. It was predicted that by 1990 at least 50 per cent of American households would be equipped with a personal computer; in fact only 15 per cent had one. For the same year it was expected that 10 per cent of American homes would be connected to a fibre-optic telephone cable, but as it turned out only less than 1 per cent enjoyed such a connection (McCarroll, 1991). Yet, telecommunications proved crucial in some other, rather unpredictable, circumstances, namely the coup defeat in the Soviet Union, in August 1991, when open telecommunications channels permitted the transmission of supporting messages from the West, through the telephone, fax, and TV. The currently enhanced and diversified telecommunications means and media are, therefore, at present in the midst of processes that have and will continue to integrate them into the webs of social, economic, and political spheres of life.

'Telecommunications and geography' sounds even more novel, but as the motto of this book demonstrates, one of the founding fathers of modern geography, Friedrich Ratzel, referred to it back in 1897. Unfortunately, however, 'despite the fundamentally geographical nature of telecommunications, . . . the topic remains underdeveloped in geography' (Abler, 1991, p.31). This lamentable situation applies foremost to North American geography, whereas in other parts of the world, notably in France, the United Kingdom, and Germany, more research and teaching efforts have been directed to the interface between geography and telecommunications.

A practical, rather than academic, guide to the topic was published in Argentina in 1953 by Pozo de Bisceglia, and a French text on the geography of telecommunications was published in 1984 by Bakis. A first reader in English appeared in 1991 (Brunn and Leinbach, 1991), and the first text on the geography of the information economy at large was published a year earlier (Hepworth, 1990). The relative dearth of publications on telecommunications by geographers has been coupled by some lack of credit for those existing geographical studies in general reviews of telecommunications research in the social sciences, such as Snow's (1988).

The primary objective of this book is to fill this lacuna by attempting to present an orderly account for a geography of telecommunications. As a young specialty within geography, most of the materials which this book draws on are empirical, since little research has been pursued along conceptual and theoretical lines. Thus, it is hoped that the discussions in the following chapters will promote and encourage more research as well as the teaching of the geography of telecommunications. It is further hoped for that the spatial perspective offered here will serve policy makers and practitioners in the field of telecommunications.

The book is divided into two parts. Part 1 is devoted to the presentation of geographical concepts for the study of telecommunications, such as the information economy, nodes, networks, flows and diffusion. Part 2 puts telecommunications within geographical contexts, namely urban, regional, national and international frameworks. The book will begin with an introductory chapter highlighting three basic areas: telecommunications, information, and the geography of telecommunications. Chapter 2 will focus on telecommunications as a material entity and spatial artefact, consisting of transmission media, networks, and nodes. Chapter 3 will then discuss the spatial dynamics of telecommunications, detailing flows and the diffusion and penetration of telecommunications innovations, as well as the interrelationships between telecommunications and transportation. Chapters 4 to 7 portray the interplay between telecommunications and geography at urban, regional, national, and international levels, in that order. The book will end with a concluding chapter, devoted to present and possible future trends.

My personal interest in the geography of telecommunications began in 1982–83, during a sabbatical year at the University of Maryland, College Park. Initially, I was intrigued by the interrelationships between telecommunications on the one hand, and spatiality and temporality at the urban level, on the other. This interest soon turned into a wider study of the impact of telecommunications on metropolitan areas. When back in Israel, my research interest turned to regional aspects of telecommunications, mainly the spatial diffusion of innovations, and the possible role of telecommunications as an agent for regional economic development. During another sabbatical year in the United States in 1989–90 my research efforts began to focus on the international level, starting with an analysis of international telephone calls around the world. With the aid of the Fulbright Foundation I was able to complete a longitudinal study of outgoing and incoming US international telephone calls, in light of the two-way movements of commodities, people, and capital. A similar study of the Israeli international system was sponsored by *Bezek*, Israel Telecommunications Co.

The preparation of this text was aided by the generous support and assistance of many institutions and individuals. *Bezek*, Israel Telecommunications Co. provided financial aid at the writing stage. I owe special thanks and gratefulness to Mr. Yitzhak Kaul, Executive Director, as well as to Dr. Valentina Gachtman, Head of the Planning Division for Strategic Human Resources, for their interest in this project. The Faculty of Social Sciences at the University of Haifa willingly provided the necessary facilities, and I wish to thank Mr. Giora Lehavi, Head of Faculty Administration, in this regard. The Department of Geography at the University of Maryland, College Park, warmly hosted me at various stages of my research on the geography of telecommunications, and I would like to acknowledge the extensive support I received from its Chairman, Prof. John Townshend, as well as from the past Chairman, Prof. Kenneth E. Corey.

Numerous members of the community of telecommunications geographers world-wide have been extremely helpful through the unselfish and timely provision of publications and pertinent information. Ms. Shoshi Mansfeld skilfully and devotedly took care of the production of the artwork. Ms. Anat Cohen served as my research assistant and graduate student during my work on Israeli international telecommunications, thoroughly analysing a complex data matrix. I was kept abreast of *The Economist* by my colleague, Prof. Stanley Waterman.

Last but not least, a special note of gratitude goes to my wife, Michal, who wholeheartedly let me delve into the writing of three books in a row, and whose comments and suggestions have been of profound importance. This book is dedicated to our daughters, Tovy, Miri, and Noga. They should know why.

Aharon Kellerman
Haifa, December 1992

PART 1
GEOGRAPHICAL CONCEPTS FOR TELECOMMUNICATIONS

Chapter 1
Introduction: Telecommunications, Information and Geography

This introductory chapter will be devoted to an elaboration of the three most basic notions addressed throughout this volume, namely telecommunications, information, and the geography of telecommunications. The following section will attempt to clarify the meaning of the term 'telecommunications', as well as to highlight the growing significance of this form of communication. In the next section, telecommunications will be put within the wider context of the information society/economy of which telecommunications is an integral part. Finally, various possible perspectives for the geography of telecommunications will be outlined, and the problems associated with a geographical analysis of telecommunications put forward.

Telecommunications: meanings and significance

'Telecommunications' may be defined along a wide spectrum, ranging from narrow to broad meanings. Traditionally, and narrowly, telecommunications refers to the plain old telephone service (POTS), namely the transmission of voice communications using analogue technologies, providing a relatively low voice quality and limited services. Widely defined, it may refer to 'all two-way communication over distance' (Abler, 1991, p.31), or to 'the transmission of signals over distances' (*International Encyclopedia of Communications*, 1989, Vol.4, p.201), and as such it may include postal services as well. Usually, the term refers to the electronic transmission of voice, data, graphic and visual information. These varied forms of communications have become possible through the integration of telephone, computer, and television technologies, as well as technological developments within the telephone industry itself, mainly through the use of optical fibres and satellites.

Telecommunications and computers

The integration of telephone and computer technologies has two major implications. First, the introduction of digital switching into the telephone industry itself, opening a whole new world of services and transitions, and second, the emergence of computer-to-computer data transmission through telephone lines. This latter development has received special attention in the literature, beginning with the coining of various names for it. The French preferred 'telematics' (Telematique), originally proposed in a report to the President of France (Nora and Minc, 1980), which has become a classic in the telecommunications literature. Telematics was more recently widely defined as 'the technical and economic phenomena arising at the intersection of the progressively merging computing and telecommunications industries' (Snow, 1988, p.171). Other, less used, suggestions for the interface between telecommunications and computers were 'compunications' (Bell, 1980), and 'intercommunications' (Falk and Abler, 1980) (which has a wider connotation, referring to other media as well).

The most widely used term for the integration of computers and telecommunications is 'information technology' (IT). It was defined 'as embracing telecommunications (telephones, switches, exchanges, cables, satellites and broadcasting), on the one hand, and computers, on the other' (Jowett and Rothwell, 1986, p.1; see also Hepworth, 1990, p.68). It is, however, necessary to note the differences between telecommunications as defined before to include data transmission, on the one hand, and the use of computers as information devices, on the other. Computers perform information *processing*, whereas telecommunications takes care of information *transmission*. Thus, the first can also be carried out in-situ without any transmission, while the second includes all forms of electronic information transmission, not merely data. From a technological viewpoint, contemporary telecommunications systems are fully computerized, but not all computer systems are fully or partially telecommunicated, and even if so, they do not always require telecommunications in order to function as information processors.

Jowett and Rothwell (1986, pp.1–4) noted several reasons for the convergence of computers and telecommunications in the post World War II era. Research efforts in American and British universities turned computers into data-processing machines, not merely business calculating devices. Electronic components that fit both telecommunications and computers were developed, and both industries relied heavily on government support for R&D efforts. Thus, both industries gained considerable experience in working on large-budget, long-term projects frequently geared to a small clientele.

Telecommunications and information technologies have been conceived as crucial as well as revolutionary elements for recent and future social and economic developments. 'Telecommunications technologies and services are increasingly seen as the central nervous system of the evolving world economy of the twenty-first century, not merely as a concomitant of future growth and welfare but as a precondition for both' (Snow, 1988, p.153). Telecommunications may be viewed as a revolution for three reasons: it allows for cheaper production of existing goods and services; it brought about the production of new goods and services, constantly becoming cheaper too; and it involved the convergence of computer and telecommunications services (Snow, 1988, p.159). This latter convergence was even considered by some as a second or third industrial revolution (OECD, 1975, p.45; Toffler, 1981). The first industrial revolution attempted to extend society's muscles, the second one attempts to extend its brain. As such it may also alter social values, culture and knowledge. Last but not least, telecommunications has done more than anything else, since the invention of money, to reduce the constraints of the physical environment on organization (Cherry, 1970).

Telecommunications, services, and information economies/societies

In the previous section telecommunications was put into the context of information technology. In the following discussion, telecommunications is discussed within the wider contexts of information economies/societies, and the even more general context of service economies (Figure 1.1).

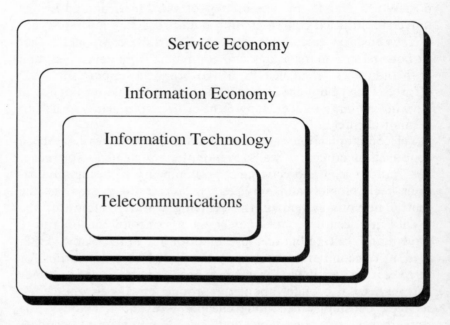

Figure 1.1 Telecommunications in information and service contexts.

Telecommunications and service economies

Early accounts of emerging service economies are usually attributed to Gottmann (1961) and Bell (1973), who proposed the widely-used term 'post-industrial society'. Beniger (1986, pp.4–5) was able, however, to count no less than 75(!) similar terms coined between 1950–84 to describe modern societal transformations. Bell (1973, p.14) proposed 'five dimensions, or components' for the post-industrial society, all of which are heavily dependent on information:

1. *Economic sector*: the change from a goods-producing to a service economy;
2. *Occupational distribution*: the pre-eminence of the professional and technical class;
3. *Axial principle*: the centrality of theoretical knowledge as the source of innovation and of policy formulation for society;
4. *Future orientation*: the control of technology and technological assessment;
5. *Decision-making*: the creation of a new 'intellectual technology'.

Service economies have evolved during the last century as a result of changes and trends in the two basic modes of economic activity, production and consumption (Figure 1.2; Kellerman, 1985). Mass production or the supply side originally depended on complementary services such as transportation and the utilities that made mass production possible, thus creating an 'intermediate demand' on these services (Daniels, 1982, p.12). Meanwhile, the evolution of mass consumption depended on distributive services, including trade and finances, insurance and real estate (FIRE) (Bell, 1973, pp.128–29). In addition, each of these two kinds of services had an important role in the development of the other mode of economic activity: the use of electricity and cars had been extremely important for the evolution of final demands or mass consumption, while trade and FIRE served as major channels for growth and distribution for mass production or supply.

At a later stage the two forces led to further development of the service economy. On the supply/production side of the economy it was the rise in productivity and the process of division of labour that led to a greater need for business services (Daniels, 1982, p.12), mainly in the form of legal and accounting services. Engel's Law of the lower elasticity of basic services increased demand for consumption of more advanced personal services such as recreation and car services (Bell, 1973, p.128; Dawson, 1979, p.71). The two processes, division of labour and

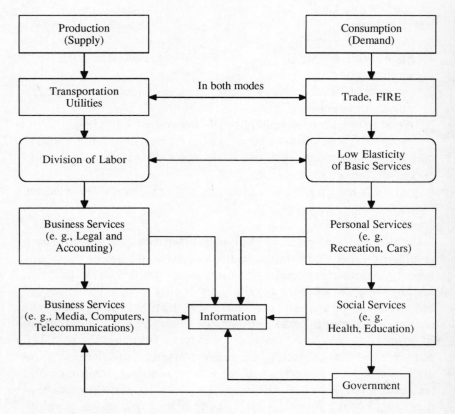

Figure 1.2 The evolution of service economies. Source: Kellerman, 1985.

Engel's Law, have been interrelated so that increased demand for more products and services caused further division of labour and vice versa.

The latest stage in the development of service economies has witnessed the rapid development of more sophisticated business services such as the increased use of communications media for advertising, and the introduction of computers and telecommunications into both production and business services. Consumption has become concerned with aspects of life quality, and hence the growing importance of social services, especially health services, education and entertainment. All these changes created pressures on governments at all levels both as service suppliers and regulators, for consumption *vis-à-vis* improved services, and for production through increased governmental involvement in research and development and in regulating competition and labour relations (Bell, 1973; Daniels, 1982).

Telecommunications and information economies

The post-industrial, service-based, economy differs from the industrial, manufacturing-based, economy in one major respect, namely in the prevailing 'strategic resource' (Bell, 1973; see also Goddard and Gillespie, 1986). In industrial economies this was finance capital, while in service economies this is knowledge, resulting from research and education. Knowledge, like its predecessor finance capital, has to be transformed in order to become economically useful. The transformation of knowledge, as well as the transformation of information into knowledge requires manipulation, processing, interpretation, exchange and transmission. These varied transformations are the responsibility of the information economy (Abler, 1977; Goddard and Gillespie, 1986). Telecommunications cannot, therefore, be equated with the information economy, since it constitutes only one component, even if crucial, of information handling, namely the electronic transmission of information and knowledge.

A geography of the information economy (such as that developed by Hepworth, 1990), would take a careful look at the sites where information processing takes place. Also a geography of telecommunications would focus on transmission processes and transmission systems for all kinds of electronic exchanges of messages. In other words, telecommunications handles information exclusively, but not all information handlings require the use of telecommunications. In telecommunications, unlike in transportation, it is difficult to separate vehicles from their cargo, or put another way, transmission software and messages are sometimes inseparable, and the telecommunications industry is highly interdependent with the service and information sectors it is bound to serve (Kellerman, 1984).

Given the growth and accompanied division of labour in modern services, it was suggested to add to the classical tertiary or service sector two additional sectors, namely the quaternary and the quinary ones (Gottmann, 1961; Bell, 1973; Abler and Adams, 1977; Abler, 1977). Originally, Gottmann (1961) referred to the 'quaternary occupations, those supplying services that require research, analysis, judgement, in brief, brainwork and responsibility' (p.580), and 'what might be called the quaternary forms of economic activity; the managerial and artistic functions, government, education, research, and the brokerage of all kinds of goods, services and securities' (p.774). Abler and Adams (1977) and Abler (1977), classified occupations along five sectors, differentiating between 'information activities' which became the quaternary sector, and the quinary sector, consisting of control functions, with government as a major quinary activity. The three traditional sectors of agriculture/mining, manufacturing, and services are based on *mode* of manipulation as a criterion, while the two additional sectors, especially the quaternary sector, single out an *object* of both production, transaction and consumption, namely, information, consisting of goods (books, computers), channels (telecommunications) and information *per se* (e.g. finances) (Kellerman, 1985).

The quaternary sector has been the only continuously growing sector in the US 1950–80, in terms of contribution to national income (Figure 1.3). The share of information occupations has been growing in countries throughout the developed world (Table 1.1; see also Hepworth 1990).

The information economy, like any other economy, has its own problems, of which three are directly relevant to telecommunications, and will be briefly mentioned here, they are, productivity, equality, and penetration. All three relate to high, yet unfulfilled, expectations. It was recently reported that while more than a third of the US GNP is generated by information, white-collar productivity is no higher than it was 30 years ago (McCarroll, 1991). Working in a 'paperless' environment probably creates its own framework and procedures, not necessarily resulting in higher productivity.

Information technology is not equally and freely available. The need to pay for information, or the 'commodification of

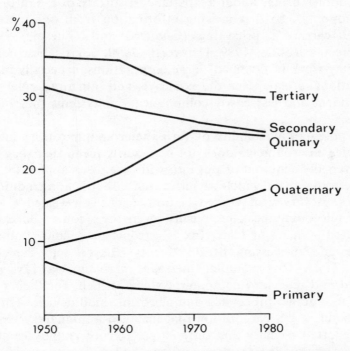

Figure 1.3 Percentage contribution of economic sectors to US national income 1950-80. Data source: Survey of Current Business. Classification source: Abler and Adams, 1977.

Table 1.1 Percentage information occupations in the labour force of selected OECD countries 1951-81

Country	1951	1961	1971	1975	1981
Australia	-	-	39.4	-	41.5
Canada	29.4	34.2	39.9	-	-
Finland	12.6	17.3	22.1	27.5	30.1
France	20.3	24.1	28.5	32.1	-
Germany	18.3	23.4	29.3	32.8	33.5
Japan	17.9	22.2	25.4	29.6	-
Sweden	26.0	28.7	32.6	34.9	36.1
United Kingdom	26.7	32.1	35.6	-	41.0
United States	30.7	34.7	41.1	-	45.8

Source: OECD, 1985; see also Hepworth, 1990, p.20.

information', brings about a 'spatial cost surface of the information economy', so that moving information from one place to another carries a price tag (Goddard and Gillespie, 1986; Gillespie and Robins, 1989; Hepworth, 1990, p.65). These conditions may work in favour of large corporations, especially in the international arena, given the extensive, and sometimes relatively independent and less costly communications systems they maintain.

The potential availability of varied telecommunications means for office environments does not necessarily mean that they will be universally acquired or put into wide use. This applies mainly to electronic mail which requires the use of non-traditional routines. A survey of United Nations agencies revealed a wide use of telecommunications means, even their convergence into integrated systems of telex, fax, electronic mail, and computer storage and processing of documents (Bikson and Schieber, 1990). Thus, for example, messages arriving via fax were scanned and read into computer systems, which, for their part, served as storage, processing and electronic mail devices. On the other hand, a US corporation that installed an e-mail system in 1988 reported recently that only 30 per cent of employees at its headquarters made use of the system, citing that 'there are too many options, and every option has suboptions. It's easier to just pick up the telephone' (McCarroll, 1991).

The geography of telecommunications

As mentioned in the preface, the geography of telecommunications may well constitute one of the less-developed branches of the discipline in terms of research accumulation and general development, and this may be attributed to several causes. First, geography in general, and economic and urban geography in particular, tend to deal with tangible artefacts, and the geographical 'inventory' of telecommunications is often either intangible or invisible (Bakis, 1981). This is especially noticeable when one compares electronic transmission media and flows of information, on the one hand, to those of people and commodities through the various modes of transportation, on the other. Whereas the latter is a heavy space-consumer, in the form of roads, parking areas, airports, seaports, railway stations, etc., the very flow of information cannot be seen at all. Cableless transmission has no visible channels, and cables of wired telecommunications means lie underground. In addition, digital telephone exchanges have become smaller and smaller in size, which is also true for computers and other communications devices. Paradoxically, however, the barely visible geography of telecommunications may well be the most extensive geography. It consists of terrestrial elements (transmission media, networks, and nodes), maritime elements (maritime cables), and space and air elements (waves, satellites).

Second, telecommunications has somehow become attached to hopes and expectations that its instantaneous nature will bring about an end to the friction of distance, or 'the tyranny of space', as well as to the disappearance of all kinds of regional and national differentiations, once the world has turned into a 'global village' (McLuhan, 1964; Toffler, 1981; see also Gillespie and Robins, 1989). Third, and related to the previous point, the study of the social effects of telecommunications, telematics, and information technology (in the general sense of the term) has carried a futuristic flavour, which may seem foreign to 'down to earth' geographers.

Fourth, there is a severe dearth in available telecommunications data, mainly for the most crucial element-flows. It is rare to find detailed data on intraurban, interurban, and interregional telephone calls, and data on international calls are usually available only in the most aggregate form of annual totals of calls placed to any other country. Thus, the healthy intercourse between theoretical argumentations and empirical studies which normally typifies emerging fields has been rather limited as far as the geography of telecommunications is concerned.

Early developments in the geography of telecommunications

Typical of an emerging field is the wrestling with verbal expressions describing new phenomena. In our case this relates to the geography of the fast flow of information, for which a myriad of names and terms has been put forward. Notable examples are: 'information flows'; 'flow societies' (Abler and Falk, 1981), 'electronic highways' (Goddard and Gillespie, 1986), 'electronic space' (Robins and Hepworth, 1988), 'space of flows' (Castells, 1985; 1989; Hepworth, 1990), and 'telegeography' (Staple, 1991).

With the exception of Ratzel's (1897) short discussion of information, an early explicit reference to communications via telephones was provided by Christaller (1933) in his classical work on central places. Christaller suggested the use of an index of telephone density as a measure of urban regional centrality, noting that:

> nothing today is as necessary or as characteristic of importance as the telephone. It is almost the symbol of whether an institution has a real central importance or only a local one. The telephone is a kind of common denominator to which all the various factors which make up the importance of a place can be reduced (p.143).

Another pioneering work was Abler's dissertation on US telephone and postal communications, followed by his exposition of a geography of communications (Abler, 1968; 1974). Gottmann (1961) elaborated on the role of information in the emerging megalopolis, while Goddard (1973) was able to point to the linkage between office locations and communications in central London. Other early works include studies published in or on communications in Sweden. Such are Tornqvist's (1968) analysis of the relationship between information flows and the location of economic activities, and Pred's (1973) and Lesko's (1990) wide-ranging reviews published in Sweden, which identified various writings from as early as the 1950s. The importance attributed to telecommunications at the time by the Swedish geographical community might have to do with the special status of telecommunications in Sweden. For years the rate of telephone penetration in Sweden has been the highest in the world, and the Swedish telephone company like its North American counterparts are the only ones that produce telecommunications equipment in addition to their provision of telecommunications services.

Basic notions for the geography of telecommunications

Two notions were originally suggested for the geographical study of telecommunications (as well as transportation) means, namely 'space-adjusting technologies' (Ackerman, 1958; see also Abler, 1975) and 'time–space convergence' (Janelle, 1968; 1991). The two notions are not identical. The first relates generally to the importance of telecommunications technologies as reducing the impact of distance, while the second 'measures the rates at which places move closer together or further away in travel or communication time' (Janelle, 1991, p.49). It may be shown quantitatively how time–space convergence has occurred through direct-dialling (in time), and through falling call-prices (in space) (Figure 1.4).

The direct geographical result of the blurring of monetary and temporal costs required in order to overcome the friction of distance may potentially be a dispersion of economic activities. However, such a dispersion process discriminates among the dispersing functions, so that existing centres of communications control may become more powerful, thus leading to further centralization of power (Innis, 1950; 1951; see also Gillespie and Robins, 1989). These contradicting impacts of telecommunications being both a decentralizing and centralizing agent may be witnessed at various geographical levels. At the metropolitan level back offices may move to suburbia, with the aid of powerful telecommunications means, but communications headquarters would prefer the central business districts (CBDs), so that city centres retain more controlling power (see also Abler and Falk, 1981). At the regional level, manufacturing may move to peripheral regions, maintaining contacts with company headquarters via telecommunications, but research and development (R&D) as well as decision-making remains in older centres. Similar processes may develop at the global level as well. Whereas the whole world may be interconnected through cheaply-priced, direct-dialling communications systems, information producing centres emerge in the world cores, as against areas of mere information consumption, located in the world peripheries. The status of peripheral regions may thus deteriorate despite seeming economic development, and this

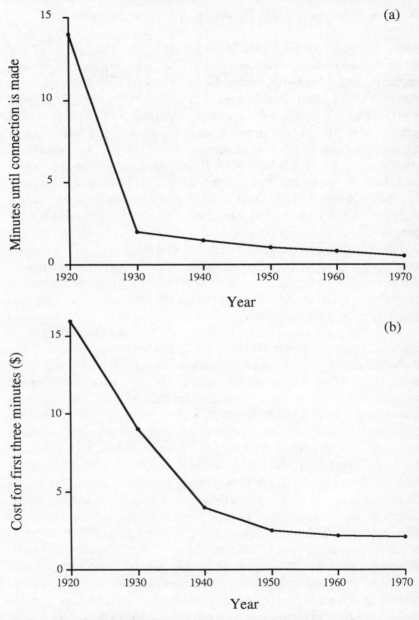

Figure 1.4 Telephone convergence between New York and San Francisco: (a) time-space convergence; (b) cost-space convergence. Source: Abler, 1975, pp.39 and 43.

may well be considered a spatial–economic divergence process, in the sense of increasing gaps between centres and peripheries at various geographical levels.

The simultaneous dispersion–concentration or convergence–divergence processes exemplify the more general two-way relationship between society and telecommunications. On the one hand, telecommunications may adjust or change the organization of space at various levels, as well as the flows of people, capital and commodities over space. On the other hand, existing social and spatial structures and contexts may shape the evolution and adoption patterns of telecommunications technologies. Thus, not every available technology may automatically be adopted by society at large, and the use and system organization of telecommunications technologies may differ from country to country. Also, the spatial expansion of cities may provide an incentive for further development of telecommunications technologies, but this is not necessarily a two-way process, so that the emergence of new telecommunications technologies does not automatically imply urban spatial expansion (Kellerman, 1989).

The classification of geographical aspects of telecommunications may be pursued along four avenues: the just mentioned study of social aspects of space adjusting by telecommunications; spatial systems; geographical levels; and interrelated sectors (Figure 1.5). One may view telecommunications as having its own geography, consisting of locational elements, such as nodes, networks and transmission media, as well as having its own dynamic elements in form of information flows, and in another way, in the diffusion of telecommunications innovations.

Telecommunications is also an integral part of spatial organization at various geographical levels. Here the role of telecommunications as a transforming agent may be emphasized, and its differential impact at various geographical levels (urban, regional, national, and international) identified. Telecommunications has been incorporated into the activities of leading sectors in space: information, banking, trade, manufacturing, services, households, and travel. Looking at the role telecommunications may play in its interaction with these sectors is a third classification,

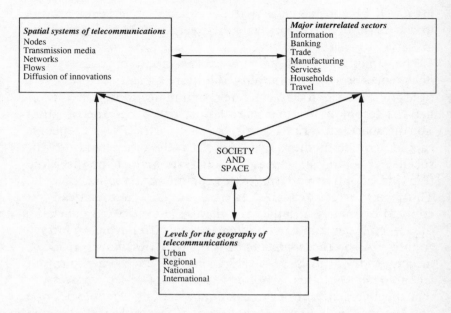

Figure 1.5 Approaches to the geography of telecommunications.

and it may too expose the contradicting geographical impacts telecommunications may have.

Conclusion

Telecommunications may be viewed as an entity by itself, as well as being an indispensable and integrated element of information technology and information and service economies. This double identity of telecommunications applies also to its geographical analysis. Thus, the geography of telecommunications *per se* has to be coupled with geographical perspectives on its roles at various spatial levels, and on its interaction and integration into social life and into the various economic sectors.

Obviously there is some overlapping among these approaches. Each of them includes elements of location and movement, as well as constraints and catalysts. The two following chapters will present the geographical elements of telecommunications, followed by expositions of its organization and functioning at various geographical levels. It will be through these two presentations that the interactions of telecommunications with other sectors will be outlined, as well.

Chapter 2
Telecommunications as Material Entity and Spatial Artefact

The previous chapter attempted to put telecommunications within various technological, economic, social and geographical contexts. In this chapter we will begin exploring telecommunications and its geography in more detail. This chapter is devoted to what Abler (1991) termed 'hardware', or the material elements of telecommunications. The first section will focus on transmission media, mainly on the telephone, the fax machine, the computer, and to a lesser extent also on television. The second section will discuss networks, concentrating on data transmission networks at local and regional levels, and on maritime cables and communications satellites at international levels. Last but not least will be telecommunications nodes, described in the third section. Here, attention will be given to smart buildings, teleports, and telecottages, their locations, characteristics and impacts.

Transmission media

This section will jointly discuss all major telecommunications transmission media, namely the telephone, the fax, and the computer. It will further focus on television technologies as far as picturephone, videoconferencing, and to a lesser extent cable television are concerned. A major emphasis will obviously be put on the telephone, since it constitutes the 'traditional' medium, being in universal use in both households and businesses, and since other means are based on its infrastructure of lines and exchanges. Three aspects will be given consideration, namely the development of telecommunications, recent trends in the telecommunications industry, and the sociospatial characteristics and impacts of telecommunications.

The development of telecommunications

Table 2.1 presents a simple chronology of innovations and development in telecommunications along the last 150 years, beginning with the introduction of the telegraph by Morse in 1837, followed by the invention of the telephone by Bell in 1876. It may well be noted that the two world wars were not marked by any innovation, though the wars, notably the second one, served as catalysts for further innovation, especially in the fields of computers and space technologies. These led the way to innovations and developments in telecommunications in the post-World War II era. There is also a marked difference between the relatively slower pace of transitions in the nineteenth century and the faster one in the twentieth century, notably in the second half of the century, a period which marks the 'telecommunications revolution'.

As is clear from the data, it has taken less time for innovations to become commercially implemented in the nineteenth century than in the twentieth. Though common wisdom could have led to an opposite conclusion, the increased complexity and sophistication of innovations in the twentieth century, have to be borne in mind (Table 2.2) (see also Abler, 1977). The commencement of transatlantic transmission of messages took decades rather than several years, following the innovation stage of communication technologies, because of its dependency on the development of long range transmission or network technologies (radio, satellites, etc.). Thus, the number of domestic telephone calls in the US exceeded the number letters already in 1939, but the same transition for the international transmission of messages occurred only in 1988. The emergence of telecommunications devices which are based on two integrated technologies (such as telex, computer-telephone and

Table 2.1 Major innovations and developments in telecommunications 1837–1988

Year	Innovation or development
1837	Telegraph demonstrated and patented
1839	Telegraph messages transmitted between Paddington and West Drayton, UK
1843	Facsimile (via telegraph lines) invented
1844	Telegraph messages transmitted between Washington and Baltimore, USA
1866	First transatlantic telegraph cable
1876	Telephone demonstrated and patented
1877	First long distance telephone line set between Bismarck's palace in Berlin and his farm at Varzin, Germany
1878	Commercial telephone switchboards and exchanges established
1884	Long-distance telephone service begins in the USA
1887	First international telephone line set between Paris and Brussels
1891	First underwater telephone cable between England and France
1896	Radio invented
1903	First non-experimental transatlantic radio conversation
1925	Television invented
1926	Transatlantic radio facsimile service begun commercially
1927	Commercial transatlantic radio-telephone service inaugurated
1928	Regular television programmes begin
1929	Colour television demonstrated
1939	Volume of domestic phone messages higher than mail in the USA
1946	Telex (telegraph exchange) developed
1948	Transistor invented
1950	Semiconductor diode invented
1951	Cable television starts
1956	First transatlantic telephone cable installed
1962	First all-electronic switches manufactured
1964	Picturephone introduced
1964	Packet switching invented permitting switching of data traffic
1965	First communications satellite put into service
1970	Combination of laser and fibre-optic technologies permitting the production of fibre-optic cables
1973	First digital data network put in service
1979	Cellular radio introduced
1982	Competition in long-distance telephone service begun in the US
1983	First long-distance fibre-optic telephone cable inaugurated
1980s	ISDN standards developed
1988	First transatlantic fibre-optic cable installed
1988	Volume of international phone messages higher than mail in the US

Sources: Beniger (1986); Dawidziuk and Preston (1981); US FCC (1988); Robinson (1977); *International Encyclopedia of Communications* (1989); MacMahon (1980); *France Information* (1984); *The Economist* (1990; 1991); Simpson (1984); Pool (1990); Kern, 1983.

Table 2.2 Time-lags between invention and implementation of telecommunications innovations

System	Innovation	Years to domestic commercial use	Years to international commercial use
Transmission media	Telegraph	2 (UK) 7 (US)	29
	Facsimile	0	83
	Telephone	2	51 (Radio) 80 (Cable) 89 (Satellite) 112 (Fibre-optic)
	Television	3	40
	Telex	0	0
Networks	Digital switching	14	–
	Computer network	9	–
	Fibre-optic	13	18

Table 2.3 Time-lags between innovation of telecommunications devices and technology integration

Technology integration	Years since telephone invention	Additional technology	Years since invention
Radio-telephone	27	Radio	7
Telex	70	Telegraph	109
Computer-telephone	97	Computer (transistor)	25
Electronic-fax	108	Computer (transistor)	36

electronic fax) has taken a relatively long time in both centuries, given the higher technological challenge involved (Table 2.3).

The geographical sources of innovation are divided between the US (telegraph, telephone, telephone switchboards, television, transistor, semiconductor, cable television, digital switching, picturephone, packet switching, communications satellites, laser and fibre-optics, cellular radio), the UK (facsimile, television, telex), and Italy (radio) (television was experimented simultaneously in the US and the UK, which is also true for telegraph,

Table 2.4 Phases of telecommunications infrastructure

Phase	Telephony	Switching systems	Transmission systems	Services
I	analogue	crossbar switches	wire/cable	telephone telegraph telex
II	analogue and digital mix	circuit and packet switches	above plus microwave, satellite, fibre-optics	above plus high-speed data comm., fax
III	digital	ISDN? virtual routing and messaging	cellular radio, cable TV	above plus TV video and higher speed (broad-band)

Source: Hart (1988).

though the latter was patented in the US by Samuel F.B. Morse). The senior share of the US (13 inventions) compared to Europe (4 inventions), is striking throughout the development of telecommunications, and strikingly during the post World War II era. It exemplifies the transition from the industrial revolution, which was headed by European nations, to the 'information revolution' dominated by the US (Kellerman, 1985).

Grouping communications technologies into parallel phases for telephonies, switching systems, transmission systems and services, results in three phases of telecommunications infrastructure (Table 2.4) (Hart, 1988; see also Goddard and Gillespie, 1986). Phase one constitutes the telecommunications infrastructure which was in use in most industrialized countries until the early 1970s. These systems permitted voice and written exchanges (via the telephone, telegraph and telex). The connection was made possible by crossbar switches based on electromagnetic relays. Wires, cables, and some microwave transmission (as of the 1950s) made for the network hardware. These systems were further characterized in the functioning of telephone companies or PTTs (government departments for postal, telephone and telegraph services) under a 'natural monopoly' typifying the provision of utilities. The cost of adding customers to large systems was small, so that new service providers could not compete with major service providers.

Phase two incorporates all the major innovations developed as of the 1960s, and implemented in the 1970s and 1980s. Exchanges could now be digitized, and many of them did, with France and the US leading the transition. This phase was referred to as a 'revolution as profound as the invention of printing' (Pool, 1990, p.7). Exchanges were computerized, permitting fast and extensive data communications with the aid of packet switches. Digital equipment permitted the simultaneous use of older analogue exchanges alongside modern digital ones. Communications of voice, video, graphics, and data has become better, clearer and faster with the introduction of new transmission channels in the form of satellites, fibre-optic cables and microwave antennas. Additional developments made it possible to use single telephone lines for several simultaneous transmissions. Furthermore, these new transmission options opened the way to the bypassing of existing wire networks, and so provided the technical background for the introduction of competition into the long-distance and international telecommunications markets, first in the US, and later in the UK, Japan, and various other countries. Another major advantage of digital systems has been the possible provision of information services over telephone lines, either through voice telephony or through computer terminals, usually termed value added networks (VANS). The British Prestel, the French Minitel, and the American Prodigy are striking examples of such information-selling systems, using subscriber computer terminals connected to the telephone system.

Phase three may become fully operational only later during this decade, depending on the advancement of ISDN (Integrated Services Digital Network). ISDN constitutes a series of technical protocols permitting the transmission of all possible forms of messages (voice, data, video, graphic) on one single telephone line, in separate or integrated forms. This line may transmit cable TV broadcasts and phone calls simultaneously, and/or serve as a picturephone or videoconferencing device integratively with data transmission. The transmission of data from a computer will not require a telephone modem, and it may be six times faster than through the best conventional telephone line and at cheaper rates. However, there are also other options for complex transmissions, such as wide bandwidth lines. By the same token, the transmission of fax messages over ISDN may be ten times faster than on a mixed analogue-digital system. ISDN can become operational on a partial base as well, namely to customers who have digital exchange service. It reaches optimal use when customers are hooked into fibre-optic cables, thus permitting wider bandwidths of transmission, which is important especially for video messages containing a high volume of digital information per picture. Currently, ISDN standards are separately developed in various countries. This provides some protection for domestic equipment suppliers. Therefore, technical solutions will have to be found in order to interconnect national ISDN systems.

Recent trends in the telecommunications industry

The discussion of ISDN attested to the transitional position of telecommunications systems in industrialized countries, being currently between phases two and three. It is thus of special interest to point to additional contemporary trends in the development and organization of transmission media. Generally, Europe is slower than North America in the transition to the digital age, with the exception of France, and the ISDN connections which are currently widely offered in the UK. European telephone systems are mostly still characterized by a mixture of the old-analogue and the new-digital switching technologies. They are further typified by a high degree of governmental involvement, with the exception of the UK and Sweden. Therefore, in most European countries telephones have to be bought from the national phone company, and until the mid-1980s licensing was required for fax machines in Italy (*The Economist*, 1991). To some degree the transitional phase is paced up by the debate on ownership of common carriers, whether they should continue as governmental PTTs (Post, Telephone and Telegraph), or whether other options should be explored and implemented. One may point to three major current trends and developments, in transmission media, in addition, or side by side with ISDN, namely VANS, videotelephony, and cellular telephones.

VANS: By 1990, nine-tenths of telephone companies' business was still derived from POTS (plain old telephone service), or simple talking (*The Economist*, 1990). Telephone companies are looking, however, for increasing shares of additional information business. It is estimated that POTS grows by about 7–8 per cent annually, while VANS expand at a rate of 25–30 per cent annually. The high growth rates for VANS relate to several areas. In the area of data transmission, phone companies look for larger volumes of data transmission over the telephone networks, as well as to sales of LANs (local area networks), which serve systems of interconnected computers. Another important area is the sale of information over the telephone. Major examples, which were mentioned already, are the French Minitel and the British Prestel systems, which provide a wide range of updated information through computer terminals hooked into the telephone systems, using videotex technology for text transmission over the telephone (Carey and Moss, 1985). Another successful example is the sale of recorded or spoken information over the telephone, by charging a special rate per minutes of use. The American 900 answering service was reported to yield $1 billion for 15,000 services in 1991 (*Ma'ariv*, 1992b).

A third avenue is the penetration of telephone companies into the production and marketing of computers, applying mainly to AT&T, America's largest telephone company, which agreed in 1982 to give up its monopoly over long-distance telecommunications in the US, in order to be able to move into the computer business, as well. However, this latter venture did not prove too successful, in a market typified by stiff competition. By 1990 the company lost $3 billion in selling computers (*The Economist*, 1990).

Videotelephony: Videotelephony relates to the ability to transmit callers' pictures simultaneously with their voices. This technology, first introduced in 1964, may be applied in two avenues, namely for the variously called videoconferencing, confravision, or teleconferencing, and for picturephones or videophones.

Goddard and Pye (1977) estimated at the time a 10 per cent reduction of interurban business travel, assuming the introduction of expensive videoconferencing systems. This might well have been a correct estimate, given the high prices of the systems at the time (up to $20,000), and the poor images and audio signals they transmitted. These problems were topped by users' complaints 'that they are prevented from swapping notes and documents and cannot ensure privacy. They grouse about having to leave their offices and miss phone calls to use the special rooms set up for videoconferencing' (McCarroll, 1991). However, it has recently been reported that British Telecom, jointly with IBM and Motorola, are currently involved in the development of a system which will permit simultaneous talking of participants, expandable up to 30 participants, at reasonable rates. In addition, a completely new system is being developed, which truly integrates telephones, computers, and television, in that participants will be seen on screen windows of desk-computers, side by side with relevant documents or tables. This equipment will be sold at less than $5000, and may, thus, represent a breakthrough in videoconferencing, with less or differently structured business travel (*Ma'ariv*, 1991).

The picturephone has been less successful even compared to the modest use of videoconferencing. It has been experimented in several countries and in various modes. Thus, the French experimented with the use the ordinary home television set combined with a telephone and a videocamera in Biarritz in 1984 (*France Information*, 1984). The system could be used to transmit room pictures or documents, or as a closed-circuit control system. Pictures could be recorded and retransmitted. However, the videophone has not yet become a mass product. The original device, introduced by AT&T back in 1964, sold for $8000. Later attempts, this time Japanese, to introduce picturephones to the American market, by Sony in 1988, and by Mitsubishi in 1991 have not proven successful (McCarroll, 1991). The product still kindles the imagination of phone companies, and in early 1992 it was reported that AT&T introduced another device selling for $1500 per unit, and British Amstrad introduced a picturephone for $860 (*Ma'ariv*, 1992a). The picturephone is of special interest, given its unique effect. Whereas the conventional telephone permitted an interactive conversation between two partners using their natural voice, the picturephone permits the same for partners' faces, simultaneously with their voices. As such, it may revolutionize interactive communications in its becoming more 'real' and diversified. It may further open new ways for a rather stratified hierarchy of communications and social interaction in both informal and work lives. There may emerge three levels of communications: a rather restricted communications form via audio conversations, an upgraded one through audiovisual exchanges, and at an even higher level face-to-face meetings will take place (Kellerman, 1984). However, the very existence of picturephones may bring about some embarrassment, since its use may sometimes be favoured only by one partner of a telephone conversation. It is not clear whether the videophone may replace to some degree physical closeness and, thus, strengthen the extended, spatially dispersed, family, or whether it could cause social damage especially in conflict situations (Short *et al.*, 1976).

Cellular telephones: Cellular telephones, introduced in 1979, have turned into a major success. They have been able to change the basic access notion of the telephone as reaching people in specific *places* to reaching people without any geographical restrictions of telephone lines, or the presence of people in specific locations (*The Economist*, 1991). As such, the cellular telephone may elevate human communications to a more instant level, turning the spatial context irrelevant. The cellular phone system permits the use of a thin slice of the wave spectrum for the simultaneous transmission of thousands of switched calls. Starting as a relatively expensive and bulky device for people on-the-road, whether executives or utility professionals, it rapidly became a routine device, smaller in size and cheaper in price.

In 1991 there were 12 million cellular telephone subscribers world-wide, two-thirds of whom in the US, UK and Japan. In Japan the number of mobile subscribers has risen at a rate of 50 per cent a year. In 1991 the US system grew by 43 per cent, and the European one by 35 per cent. In Sweden the penetration of this technology is the deepest, with 7 per cent of the population subscribed to the system in 1991 (*The Economist*, 1992). The technology is also used for commercial telephone service from planes. A universal adoption of this technology may turn the existing extensive wire and cable systems into auxiliary ones at least as far as voice communications is concerned.

Sociospatial characteristics and impacts of telecommunications

The sociospatial characteristics and impacts of telecommunications are typified by their contradictory and dualistic nature (Figure 2.1). Taylor (1983) pointed to the unique dualistic nature of demand for telecommunications services, its consisting simultaneously of demand for two services, namely access to and use of the telecommunications system. One may view these two, tied-together, demand elements, as representing the more passive side of telecommunications, as means for the receiving of messages (access) (see also Ball, 1968, p.63), on the one hand, and its more active role permitting the sending of messages (use), on the other. By the same token, the social and spatial aspects of access seem, though not exclusively, to pertain more strongly to the micro level of the individual user of the telephone. On the other hand, the aspects concerning the use of telecommunications lean more to the macro level of wider social systems, as well as to various telecommunications means, not just the telephone.

As far as access is concerned 'the telephone is an *irresistible intruder* in time and place' (McLuhan, 1964, p.271), and 'electronic messages [however] do not make social entrances; they steal into places like thieves in the night' (Meyrowitz, 1985, p.117). This intrusion seems to constitute the most basic sociospatial aspect of access to the telecommunications system, and it carries along a variety of social, temporal and geographical effects. It is no wonder, therefore, that the early diffusion of the telephone encountered objection to its social intrusion and accessibility (Marvin, 1988, pp.102–8).

The most striking effect of telecommunications at the social level is a democratization process, which permits an access of all household members to an interactive information system.

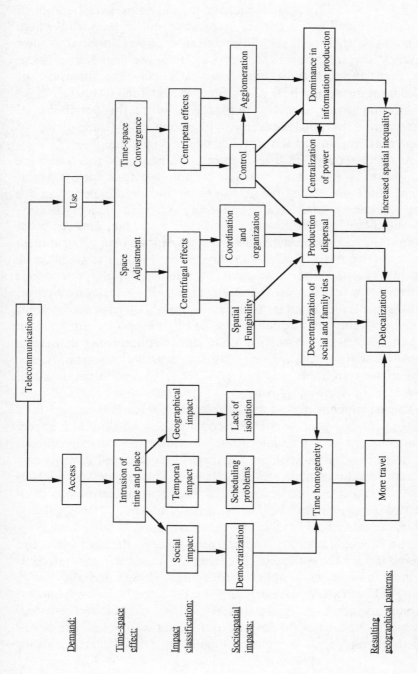

Figure 2.1 Sociospatial impacts of telecommunications.

This free access had an impact on the changing social status of women, as well as on the status of children, in permitting them to maintain their own social ties and in exposing children to their parents' social contacts (Meyrowitz, 1985; see also Kern, 1983). At the temporal level, the potentially constant intrusion of telecommunications into one's time, meant that ultimately tight scheduling became next to impossible, given the possibility of interruption by and engagement in phone calls (Ball, 1968), at least until the introduction of answering machines.

At the geographical level, the place-intrusion of the telephone has meant a direct and real-time communications form without presence of the calling party. Messages and even experiences are not blocked by house-walls or by any other physical partitions. Furthermore, telephones are universally available, and for most people they must be available everywhere, so that geographical isolation is almost impossible (Meyrowitz, 1985). People must be 'within reach' rather than 'out of touch'. Cellular telephones amplify this process, so that soon a 'person–place convergence' will be achieved, similar to the time–space convergence brought about by the traditional 'place-fixed' telephone. In modern urban life, the place-intrusion of the telephone has assisted a form of spatial isolation, namely single-person households, for which the telephone may relieve feelings of loneliness, anxiety, and lack of security (Aronson, 1971; Gottmann, 1977).

The permanent access to the telecommunications system may yield increased processes of homogenization of time and space. The constant access to telecommunications may destroy notions of before and after, eliminate temporal hierarchies, and making times interchangeable (Raulet, 1991). This process, which is related to various other facets of capitalist societies has been termed elsewhere as the 'spatialization of time' (Gross, 1981; 1985; see also Kellerman, 1989; 1991a).

The homogenization of space may emerge through increased travel induced by telephone contacts. Access to the telecommunications system implies information flows, and these may bring about an imbalance and tension 'between individuals' enhanced informational status and their still limited physical mobility. New rules of physical movement and access are developed, therefore, as a means for aligning spatial configurations with the new patterns of information flow' (Meyrowitz, 1985, p.180). More extensive travelling may be viewed as 'positive' dynamics, exposing people to new places, people and experiences, but the constant movement of people may tend to homogenize the experience and even the physical lookings of space, a process which is termed here as 'delocalization' (Raulet, 1991).

Turning now to the sociospatial aspects of the other side of demand for telecommunications, namely its use, the two most related sociospatial characteristics are space adjustment and time–space convergence, which were mentioned in the previous chapter. Space and distance are perceived and practically used as shrinking dimensions, with the emergence of high-speed, high-quality and reasonably priced, long-distance communications (see Abler, 1975; 1977). The sociospatial patterns and processes that may emerge in response to space adjustment and time–space convergence might be of a seemingly contradictory nature, namely centrifugal and centripetal effects (see Abler and Falk, 1981). As will be shown, these contradictory effects are interrelated.

The use of telecommunications involves centrifugal aspects in lieu of several of its merits. The interactive nature of communications devices, mainly the telephone, turns them into coordinating and organizational means, which permit the spatial expansion, separation, extension, and dispersal of economic activity (Pool, 1977, p.116; Abler and Falk, 1981). Furthermore, the telephone has made space fungible, in the social sense, so that everybody can communicate with everybody else (Gottmann, 1977). Raulet (1991) pointed to a similar effect from an economic perspective, referring to the interchangability of places, when they are equally accessible, so that production may diffuse geographically. These centrifugal effects may thus potentially yield a decentralization of both social ties and production. Socially, the telephone allowed for continued close contacts among spatially dispersing family members and friends, or an extension of one's 'psychological neighborhood' (Aronson, 1971; Ball, 1968). Raulet (1991) relates this increased level of sociospatial mobility to the 'deterritorialization of producers', no longer tied to landed fortunes but rather to financial ones, a process that Marx viewed as the origin of capitalism. The decentralization of production may in principle take several forms and various geographical scales. Thus, offices may tend to suburbanize, whereas manufacturing may move into peripheral regions. In the second part of the book we will see that these potential dispersion processes are highly conditioned by circumstances in both cores and peripheries.

The decentralization of production and residences could evolve into 'delocalization', namely that ties of dispersing persons to the local social sphere may weaken, once telecommunications permits close and instant contacts with people and activities remotely located in a previous place of residence. This kind of 'spatialization by communication' implies a certain level of spatial homogenization, in which the experience of deep attachment to place may be replaced with a relative position in space (Raulet, 1991), and the 'specialness of place and time' may be destroyed (Meyrowitz, 1985, p.125). Increased travel, stemming from access to telecommunications, obviously amplifies this process of dissociation between physical place and social 'place' or status (Meyrowitz, 1985, p.115).

The 'flattening' and homogenization of temporal and spatial experiences which were attributed to the access and use of telecommunications are coupled with a deterioration in the permanent recording of feelings, experiences, and thought through letter writing, and the telephone, fax, and computer serve as substitutes for thoughtful written communications. The telephone enjoys advantages over letter writing. A telephone conversation may be more spontaneous in nature, and it may accentuate events and thoughts pertaining to the present. It may further avoid the delay in sharing which is unavoidable in letter correspondence. However, letter writing involves a commitment of words to paper, namely the understanding that a letter creates a record. Also, letters create some intimacy, and some temporal distance, in the writer's knowing that the addressee will receive the letter only after a while (Meyrowitz, 1985; Rubinstein, 1991; Kern, 1983).

The centripetal effects of telecommunications relate to the strengthening of economic cores when telecommunications means are enhanced. This applies to urban cores as well as to countries and continents. The production and processing of information, whether in the form of electronic and printed media, or in the form of company and the bank headquarters, as well as in the form of capital markets, tends to agglomerate in major urban centres, located in industrialized countries. This concentration implies growth in the controlling power of such centres. The dispersion of industrial production and routine services, which represents the production power accumulated in the cores, may bring about an even higher power concentration in the cores as controlling foci, which is enhanced by the increased production of information. Development in outer areas may be offset by growth in the cores, and spatial inequality may even increase, as a result of the simultaneous and interrelated centrifugal and centripetal effects of enhanced telecommunications systems.

Networks

'Communications processing is a complex activity that is normally transparent to, or hidden from, the user' (Hepworth, 1990, p.48). This hidden dimension of telecommunications, and the technical nature of network structures could be counted as major reasons for an almost lack of interest among geographers in the topology, structure and design of telecommunications networks. Once a diversified and sophisticated system is available to a user there is no difference in costs for routing of messages one way or another. With ISDN widely installed, it might well be that charging for information transmission will be by volume of transmission of digital information rather than by distance (Goddard and Gillespie, 1986). Still, the spatial organization of the various telecommunications networks has its own patterns and logic, and these might be of some interest within the general framework of the geography of telecommunications.

Networks and exchanges have several functions: 'establishing user/network interfaces, message switching and routing, polling and addressing, speed and code conversion, message formatting, buffering, queue management, error checking, diagnostics and record-keeping' (Hepworth, 1990, p.48). Currently, there are several network technologies in use: cables in the form of copper wires, coaxial cables, fibre-optic cables; and microwaves and satellites for cableless transmission. These technologies vary in price and transmission capability. Also, the unavailability of certain transmission networks, such as fibre-optics, may lead to regional-economic backwardness (Goddard and Gillespie, 1986). Wires are simple, cheap and sturdy, but are limited in their signal transmission ability, since they were meant to carry voice messages only. They can carry low speed data too, especially to short distances. Coaxial cables, on the other hand, can carry even broadband television pictures, but they are expensive. Fibre-optic cables provide the best transmission among all network technologies for voice, data and video signals. Despite the tremendous decline in their price, the rewiring of existing extensive copper networks is going to be expensive. The estimated cost is at least $150 billion for the US. Microwaves may too carry large bandwidths, and less expensively, but their use involves interception and atmospheric problems. These and other problems, such as difficulties with interactive conversations, apply to satellites as well, though their operation is cheap (Abler, 1991). In the following paragraphs, various geographical aspects of cable networks, satellites, and maritime cable systems will be presented.

Cable networks

Telephone networks (as well as computer and cable-television networks) may be structured in principle along one of two principles of topology. Maximum connectivity may be achieved if all subscribers are linked to each other without any switching facilities, in a kind of a 'mesh' form (Figure 2.2). This pattern may imply savings in switching exchanges but is extremely expensive in network routes. As Abler (1974) showed the number of links in such a system would reach

$$\frac{N(N-1)}{2}$$

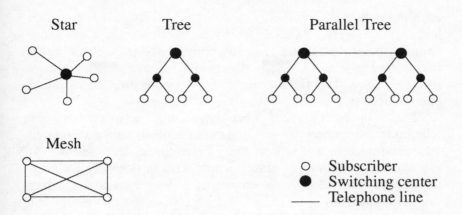

Figure 2.2 Basic topologies for spatial networks of telephone systems.

where N is the number of subscriber points. On the other hand, the use of switched systems, requires only N-1 links, to which the costs of construction and operation of switching systems have to be added. A simple switched system would, therefore, take a 'star' form. Decreasing switching costs have brought about hierarchical structures in the forms of a 'tree' or even in form of interconnected 'parallel trees'. Further declines in switching costs, as well as the introduction of digital computerized switching systems resulted in complex and mixed forms, so that contemporary communications processing is both spatially distributed and hierarchically organized (Hepworth, 1990).

Hepworth (1990, pp.39–69) went into detail regarding the network architecture and communications protocols for computer systems. Basically, data processing which is divided among several computers may be spatially distributed along patterns similar to those of telephone networks. Computer systems may use the limited-bandwidth and slow copper-wire telephone system, or they may use leased, large bandwidth, dedicated point-to-point lines permanently connecting two or more computers. Star-like networks typify centralized systems, with a mainframe computer connected to local terminals, such as a chain of retail stores. Tree-formed networks typify multi-locational organizations with a hierarchical structure, for example a bank's national/regional/local system of branches and headquarters. Parallel trees may be used in even more complex organizational and geographical systems, such as multinational corporations. Mesh networking, where each computer has the same functional status may be useful for product distribution systems.

The rapid evolution of fibre-optic cable networks by new telephone companies in the American and British deregulated telecommunications markets has been aided by the use of existing rights-of-way of transportation networks. These rights-of-way are controlled by governments, they connect major urban markets, and they are cleared and accessible. Examples are the MCI line along the AMTRAK railway between New York and Washington, the rights-of-way provided by the City of Chicago in century-old coal tunnels, and the use made by Mercury Communications of the old and unused rights-of-way of the hydraulic system of London (Moss, 1991). In Britain, BRT (British Rail Telecommunications) could potentially turn into a full-scale telecommunications company (Hepworth and Ducatel, 1992, pp.13–14).

Microwave and satellite communications

Microwave communications has been used as an integral part of national telephone networks, i.e. the French one (Figure 2.3). As mentioned earlier, this technology constitutes the backbone of cellular telephone systems. Private microwave communications systems established by corporations with heavy demands for intraorganizational telecommunications proliferated in the US already back in the 1950s, when this option was permitted by the Federal Communications Commission (FCC). This technology assisted the emergence of competitive long-distance phone companies, following a court decision in 1969, because of the low investments required for network infrastructure compared to cables (Langdale, 1983).

Communications satellites located in space, perform tasks similar to those taken care of by earth microwave relay stations, namely the transmission of digital signals for voice, data, and

Telecommunications as material entity & spatial artefact 39

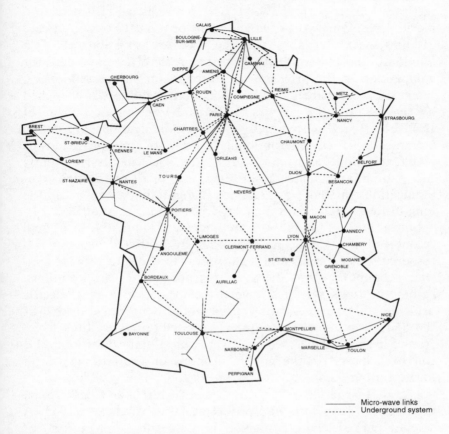

Figure 2.3 Major telephone links in France, 1984. Source: *France Information*, 1984.

video communications. Satellites require a pair of frequencies, one for uplink from the ground to the satellite, and another one for downlink from the satellite to the ground in order to avoid interference between the two in the satellite. Transponders in the satellites transform uplink signals to downlink ones. Ground dish antennas are connected to terrestrial cable systems which transmit the messages to and from connecting subscribers. Satellites permit, therefore, wireless long-distance communications, which may be used for both domestic and international communications. The wireless operation of communications satellites further permits the bypassing of existing terrestrial and maritime cable networks, which might be a useful option in case of bottlenecks notably for the fast transmission of data (Goddard and Gillespie, 1986), or in order to accommodate for competition in the provision of telecommunications services.

The first experimental communications satellite was launched by the US in 1962. It was followed by the establishment of INTELSAT (International Telecommunications Satellite Organization), an international agency for the provision of global telephone and television services via satellites, currently 115 nations are members. In 1965, the first commercial communications satellite was launched inaugurating a new era in the long-distance transmission of all types of information.

Space has an image of an endlessly abundant locational resource, but in fact it is a profoundly restricted geographical resource for communications purposes. In order to provide constant communications between two or more earth stations, communications satellites have to be located in a very specific orbit, in which their orbiting speed would be identical to that of the daily rotation of the globe. The satellites would thus seem, from a terrestrial perspective, as having a static location. This can be achieved in the geosynchronous or geostationary orbit located around the equator at an altitude of 36,000 km above sea level. This circle, or corridor, is also called the Clark Circle, after Sir Arthur Clark who proposed its use already in 1945. A satellite parked in this orbit may be visible from 43 per cent of the earth surface, so that three communications satellites positioned at appropriate distances from each other are required to cover the whole globe, with the obvious exception of the two poles. The full geostationary circle is about 260,000 km long, so that a one degree segment is about 700 km in length. There are various opinions on the optimal distance which is required between satellites, but like on the earth surface, there are more favourable locations than others, given the unevenness of the geographical spread of demand for telecommunications services. For example, a satellite positioned above the Indian Ocean can beam simultaneously to the UK and Japan. Points over the Atlantic are also in high demand because of the intense telecommunications traffic between North America and Europe. By the same token, the more favourable launching sites are close to the equator, where the rotating speed of the globe is fastest, so that less fuel and control equipment is needed to put the satellite into its space-position, so permitting heavier communications payload.

The limited spatial-locational resources for the communications satellite system have, therefore, to be internationally allocated into 'slots' or 'parking spaces'. Another limited resource for communications satellites is the wave transmission spectrum. The preferred one is the *C band*, which has become so intensely used that another band, the *Ku band*, has become operative. The latter band is more vulnerable to atmospheric disturbances, and it requires more sophisticated earth equipment. The ITU (International Telecommunications Union), and its WARCs (World Administrative Radio Conferences) are in charge of spectrum and space slot allocations among nations, on a non-compulsory basis (Figures 2.4 and 2.5). This international allocation of restricted resources has its own geopolitical problems. Is the geostationary orbit a property of the equatorial countries above which it extends, or is it a 'common heritage of mankind'? Should the slots be allocated on a demand and first-come first-served basis, thus favouring the rich nations, or should slots be preserved for developing nations? (Pool, 1990, pp.30–31; Snow, 1988, pp.166–77; Adrian-Bueckling, 1982).

The US has traditionally constituted a leader in communications satellites. Between 1962 and 1974, when the Soviets successfully launched Stationar 1, the US was the only nation equipped with the technology to place satellites in the geostationary orbit (Pool, 1990, p.30). Americans have, therefore, relied heavily on satellites for both domestic and international telecommunications. About one-third of the world's transponders and about one-fifth of the world's satellites are American, competing only with the international INTELSAT satellites which contain another third of the available transponders (Table 2.5). On average each satellite contains about 11.5 transponders, but INTELSAT satellites are usually larger and Soviet ones the smallest. Until some ten years ago, the American COMSAT company placed all US satellites. Since then, however, competition was introduced, initiated

42 Telecommunications & geography

Orbit	Inc.	Satellite Name	Operator/Country	Date	Bands	Pol.
55.5 W	*	Inmarsat II F4	Inmarsat	1992	C5/L	LP
53 W	±0.1°	INTELSAT 513	INTELSAT	1988	C1/K1-4	CP/LP
50 W	*	INTELSAT 506	INTELSAT	1983	C5/K3	CP/LP
45.5 W	±0.1	PAS-1	Alpha Lyracom	1988	C1/K1	LP
43 W	–	PAS-2	Alpha Lyracom	1993	C1/K1	LP
41 W	±0.1°	TDRS-EAST	Columbia [U.S.]	1990	C1/K	LP
39.5 W	–	PAS-3	Alpha Lyracom	1994	C1/K1	LP
34.5 W	–	INTELSAT 603	INTELSAT	1992	C2/K	CP/LP
27.5 W	±0.1°	INTELSAT 601	INTELSAT	1992	C2/K	CP/LP
26.0 W	*	Inmarsat II F2	Inmarsat	1991	C5/L	LP
24.5 W	±0.1°	INTELSAT 605	INTELSAT	1991	C2/K	CP/LP
25.0 W	*	Stat. 8/Raduga 23	CIS	1989	C3	RH
21.5 W	±1.1°	INTELSAT 502	INTELSAT	1980	C1/K3	CP/LP
18 W	±0.1°	INTELSAT 515	INTELSAT	1989	C1/K3-4	CP/LP
14 W	*	Stat. 4/Gorizont 15	CIS/Intersputnik	1986	C3/K5	RH
11 W	*	Stat. 11/Gorizont 11	CIS	1985	C3/K5	RH
8 W	±0.1°	Telecom II-A	France	1992	C1/K4	LH/LP
5 W	±0.1°	Telecom II-B	France	1992	C1/K4	LH/LP
1 W	*	INTELSAT 512	INTELSAT	1985	C1/K3	CP/LP
3 E	±0.1°	Telecom I-C	France	1988	C1/K3	LH/L
15 E	–	Amos 1 & 2	Israel	1995	C1/K3	LP
19 E	–	Arabsat II F1	Arab Sat. Org.	1994	C1/K/S	CP/LP
26 E	–	Arabsat II F2	Arab Sat. Org.	1995	C1/K/S	CP/LP
31 E	–	Arabsat 1-C	Arab Sat Org.	1992	C1/S	CP
35 E	*	Stat. 2/Raduga 22	CIS	1988	C4	RH
40 E	*	Stat. 12/Gorizont 12	CIS	1986	C3/K5	RH
45 E	*	Stat. 9/Raduga 19	CIS	1986	C4	RH
49 E	*	Stat. 24/Raduga 1-1	CIS	1990	C3 or C4	RH
53 E	*	Stat. 5/Gorizont 17	CIS	1989	C3/K5	RH
57 E	*	INTELSAT 507	INTELSAT	1983	C1/K3	CP/LP
60 E	±0.1°	INTELSAT 604	INTELSAT	1990	C2/K	CP/LP
63 E	±0.1°	INTELSAT 602	INTELSAT	1981	C2/K	CP/LP
66 E	*	INTELSAT 505	INTELSAT	1982	C1/K3	CP/LP
66.5 E	±0.1°	Inmarsat II F1	Inmarsat	1990	C5/L	CP
68 E	–	PAS-6	Alpha Lyracom	1995	C/K	–
70 E	*	Stat. 20/Raduga 25	CIS	1990	C4	RH
70 E	–	Unicom F2	Unicom [U.S.]	1997	C1/K	LP
72 E	–	PAS-7	Alpha Lyracom	1995	C1/K	–
74 E	–	Insat II-A	ISRO [India]	1994	C1/S	LP
77.5 E	–	Asiasat 2	Asiasat [Hong Kong]	1995	C1/K	LP
78.5 E	–	Thaicom A2	SCC [Thailand]	1994	C1/K	LP
80 E	*	Stat. 13/Gorizont 16	CIS/Intersputnik	1988	C3/K5	RH
83 E	±0.1°	Insat 1-D/II-C	ISRO [India]	1990/93	C1/S	LP
85 E	*	Stat.3/Raduga 20	CIS	1987	C4	RH
87.5 E	±0.1°	Chinasat 1	China	1988	C1	LP
90 E	*	Stat. 6/Gorizont 20	CIS	1990	C3/K5	RH
91.5 E	*	INTELSAT 501	INTELSAT	1981	C1/K3	CP/LP
93.5 E	–	Insat II-B	ISRO [India]	1992	C1/S	LP
96.5 E	*	Stat. 14/Gorizont 19	CIS	1989	C3/K5	RH
101.5 E	–	Thaicom A1	SCC [Thailand]	1994	C1/K	LP
103 E	*	Stat. 21/Gorizont 12	CIS	1986	C3	RH
105.5 E	±0.1°	Asiasat 1	Asiasat [Hong Kong]	1989	C1	LP
108 E	±0.1°	Palapa B2R/C1	Telekom [Indonesia]	1990/95	C1	LP
110.5 E	±0.1°	Chinasat 2	China	1988	C1	LP
113 E	±0.1°	Palapa B2P/C2	Telekom [Indonesia]	1987/95	C1	LP
118 E	±0.1°	Palapa B4	Telekom [Indonesia]	1992	C1	LP

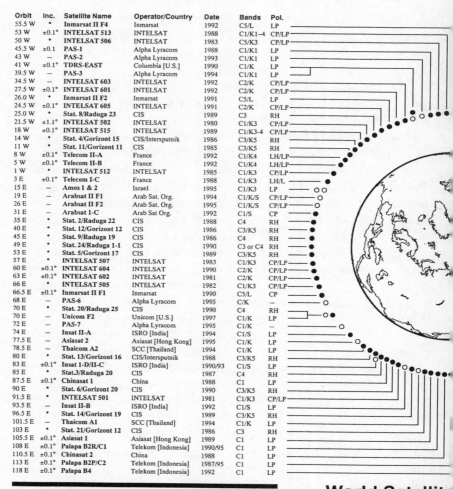

© 1992 MLE INC
All Rights Reserved
Post Office Box 159,
Winter Beach, FL 32971
Tel: (305) 767-4687
Fax: (305) 767-6067

World Satellite
C-Band Satellite

U =	700 to 730 MHz	L =	1.5 to 1.65 GHz
C2 =	3.65 to 4.2 GHz	C3 =	3.65 to 3.95 GHz
K1 =	11.7 to 12.2 GHz	K2 =	12.2 to 12.7 GHz
K5 =	11.52 to 11.56 GHz	K6 =	11.7 to 12.5 GHz
Ka =	19.1 to 20.2 GHz	CP =	Circular Polarization
VP =	Vertical only	HP =	Horizontal only

Figure 2.4 The allocation of the geostationary orbit for C-band satellites, 1992. Source: Long, 1992.

Telecommunications as material entity & spatial artefact 43

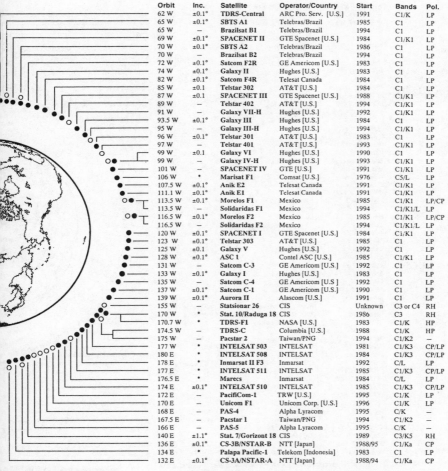

Orbit	Inc.	Satellite	Operator/Country	Start	Bands	Pol.
62 W	±0.1°	TDRS-Central	ARC Pro. Serv. [U.S.]	1991	C1/K	LP
65 W	±0.1°	SBTS A1	Telebras/Brazil	1985	C1	LP
65 W	–	Brazilsat B1	Telebras/Brazil	1994	C1	LP
69 W	±0.1°	SPACENET II	GTE Spacenet [U.S.]	1984	C1/K1	LP
70 W	±0.1°	SBTS A2	Telebras/Brazil	1986	C1	LP
70 W	–	Brazilsat B2	Telebras/Brazil	1994	C1	LP
72 W	±0.1°	Satcom F2R	GE Americom [U.S.]	1983	C1	LP
74 W	±0.1°	Galaxy II	Hughes [U.S.]	1983	C1	LP
82 W	±0.1°	Satcom F4R	Telesat Canada	1984	C1	LP
85 W	±0.1	Telstar 302	AT&T [U.S.]	1984	C1	LP
87 W	±0.1	SPACENET III	GTE Spacenet [U.S.]	1988	C1/K1	LP
89 W	–	Telstar 402	AT&T [U.S.]	1994	C1/K1	LP
91 W	–	Galaxy VII-H	Hughes [U.S.]	1992	C1/K1	LP
93.5 W	±0.1°	Galaxy III	Hughes [U.S.]	1984	C1	LP
95 W	–	Galaxy III-H	Hughes [U.S.]	1994	C1/K1	LP
96 W	±0.1°	Telstar 301	AT&T [U.S.]	1983	C1	LP
97 W	–	Telstar 401	AT&T [U.S.]	1993	C1/K1	LP
99 W	±0.1	Galaxy VI	Hughes [U.S.]	1990	C1	LP
99 W	–	Galaxy IV-H	Hughes [U.S.]	1993	C1/K1	LP
101 W	–	SPACENET IV	GTE [U.S.]	1991	C1/K1	LP
106 W	*	Marisat F1	Comsat [U.S.]	1976	C5/L	LP
107.5 W	±0.1°	Anik E2	Telesat Canada	1991	C1/K1	LP
111.1 W	±0.1°	Anik E1	Telesat Canada	1991	C1/K1	LP
113.5 W	±0.1°	Morelos F1	Mexico	1985	C1/K1	LP/CP
113.5 W	–	Solidaridas F1	Mexico	1994	C1/K1/L	LP
116.5 W	±0.1°	Morelos F2	Mexico	1985	C1/K1	LP/CP
116.5 W	–	Solidaridas F2	Mexico	1994	C1/K1/L	LP
120 W	±0.1°	SPACENET I	GTE Spacenet [U.S.]	1984	C1/K1	LP
123 W	±0.1°	Telstar 303	AT&T [U.S.]	1985	C1	LP
125 W	±0.1	Galaxy V	Hughes [U.S.]	1992	C1	LP
128 W	±0.1°	ASC 1	Contel ASC [U.S.]	1985	C1/K1	LP
131 W	–	Satcom C-3	GE Americom [U.S.]	1992	C1	LP
133 W	±0.1°	Galaxy I	Hughes [U.S.]	1983	C1	LP
135 W	–	Satcom C-4	GE Americom [U.S]	1992	C1	LP
137 W	–	Satcom C-1	GE Americom [U.S.]	1990	C1	LP
139 W	±0.1°	Aurora II	Alascom [U.S.]	1991	C1	LP
155 W	–	Statsionar 26	CIS	Unknown	C3 or C4	RH
170 W	*	Stat. 10/Raduga 18	CIS	1986	C3	RH
170.7 W	*	TDRS-F1	NASA [U.S.]	1983	C1/K	HP
174.5 W	–	TDRS-C	Columbia [U.S.]	1988	C1/K	HP
175 W	*	Pacstar 2	Taiwan/PNG	1994	C1/K2	–
177 W	*	INTELSAT 503	INTELSAT	1981	C1/K3	CP/LP
180 E	*	INTELSAT 508	INTELSAT	1984	C1/K3	CP/LP
178 E	*	Inmarsat II F3	Inmarsat	1992	C/L	LP
177 E	*	INTELSAT 511	INTELSAT	1985	C1/K3	CP/LP
176.5 E	*	Marecs	Inmarsat	1984	C/L	LP
174 E	±0.1°	INTELSAT 510	INTELSAT	1985	C1/K3	CP/LP
172 E	–	PacifiCom-1	TRW [U.S.]	1995	C1/K	LP
170 E	–	Unicom F1	Unicom Corp. [U.S.]	1996	C1/K	LP
168 E	–	PAS-4	Alpha Lyracom	1995	C/K	–
167.5 E	*	Pacstar 1	Taiwan/PNG	1994	C1/K2	–
166 E	–	PAS-5	Alpha Lyracom	1995	C/K	–
140 E	±1.1°	Stat. 7/Gorizont 18	CIS	1989	C3/K5	RH
136 E	±0.1°	CS-3B/NSTAR-B	NTT [Japan]	1988/95	C1/Ka	CP
134 E	*	Palapa Pacific-1	Telekom [Indonesia]	1983	C1	LP
132 E	±0.1°	CS-3A/NSTAR-A	NTT [Japan]	1988/94	C1/Ka	CP

A L M A N A C
Wall Chart

S =	2.5 to 2.6 GHz	C1 =	3.7 to 4.2 GHz
C4 =	3.4 to 3.675 GHz	C5 =	4.160 to 4.198 GHz
K3 =	10.95 to 11.7 GHz	K4 =	12.5 to 12.75 GHz
K7 =	10.75 to 10.95 GHz	K8 =	11.7 to 11.9 GHz
RH =	Right Hand Circular	LP =	Horizontal/Vertical
* =	Inclined Orbit		

All the world's satellites, explained, illustrated, footprinted, indexed and analysed in detail!
(Future INTELSAT deployment plans can be found on page 156 of the new third edition of the World Satellite A L M A N A C)

44 Telecommunications & geography

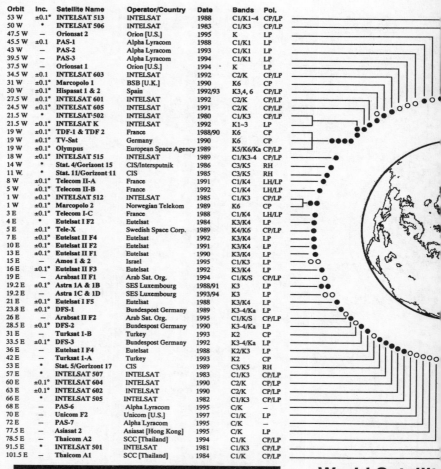

Orbit	Inc.	Satellite Name	Operator/Country	Date	Bands	Pol.
53 W	±0.1°	INTELSAT 513	INTELSAT	1988	C1/K1–4	CP/LP
50 W	*	INTELSAT 506	INTELSAT	1983	C1/K3	CP/LP
47.5 W	–	Orionsat 2	Orion [U.S.]	1995	K	LP
45.5 W	±0.1	PAS-1	Alpha Lyracom	1988	C1/K1	LP
43 W	–	PAS-2	Alpha Lyracom	1993	C1/K1	LP
39.5 W	–	PAS-3	Alpha Lyracom	1994	C1/K1	LP
37.5 W	–	Orionsat 1	Orion [U.S.]	1994	K	LP
34.5 W	±0.1	INTELSAT 603	INTELSAT	1992	C2/K	CP/LP
31 W	±0.1°	Marcopolo 1	BSB [U.K.]	1990	K6	CP
30 W	±0.1°	Hispasat 1 & 2	Spain	1992/93	K3,4, 6	CP/LP
27.5 W	±0.1°	INTELSAT 601	INTELSAT	1992	C2/K	CP/LP
24.5 W	±0.1°	INTELSAT 605	INTELSAT	1991	C2/K	CP/LP
21.5 W	*	INTELSAT 502	INTELSAT	1980	C1/K3	CP/LP
21.5 W	±0.1°	INTELSAT K	INTELSAT	1992	K1–3	LP
19 W	±0.1°	TDF-1 & TDF 2	France	1988/90	K6	CP
19 W	±0.1°	TV-Sat	Germany	1990	K6	CP
19 W	±0.1°	Olympus	European Space Agency	1989	K5/K6/Ka	CP/LP
18 W	±0.1°	INTELSAT 515	INTELSAT	1989	C1/K3-4	CP/LP
14 W	*	Stat. 4/Gorizont 15	CIS/Intersputnik	1986	C3/K5	RH
11 W.	*	Stat. 11/Gorizont 11	CIS	1985	C3/K5	RH
8 W	±0.1°	Telecom II-A	France	1991	C1/K4	LH/LP
5 W	±0.1°	Telecom II-B	France	1992	C1/K4	LH/LP
1 W	±0.1°	INTELSAT 512	INTELSAT	1985	C1/K3	CP/LP
1 W	±0.1°	Marcopolo 2	Norwegian Telekom	1989	K6	CP
3 E	±0.1°	Telecom I-C	France	1988	C1/K4	LH/LP
4 E	*	Eutelsat I F2	Eutelsat	1984	K3/K4	LP
5 E	±0.1°	Tele-X	Swedish Space Corp.	1989	K4/K6	CP/LP
7 E	±0.1°	Eutelsat II F4	Eutelsat	1992	K3/K4	LP
10 E	±0.1°	Eutelsat II F2	Eutelsat	1991	K3/K4	LP
13 E	±0.1°	Eutelsat II F1	Eutelsat	1990	K3/K4	LP
15 E	–	Amos 1 & 2	Israel	1995	C1/K3	LP
16 E	±0.1°	Eutelsat II F3	Eutelsat	1992	K3/K4	LP
19 E	–	Arabsat II F1	Arab Sat. Org.	1994	C1/K/S	CP/LP
19.2 E	±0.1°	Astra 1A & 1B	SES Luxembourg	1988/91	K3	LP
19.2 E	–	Astra 1C & 1D	SES Luxembourg	1993/94	K3	LP
21 E	±0.1°	Eutelsat I F5	Eutelsat	1988	K3/K4	LP
23.8 E	±0.1°	DFS-1	Bundespost Germany	1989	K3-4/Ka	LP
26 E	–	Arabsat II F2	Arab Sat. Org.	1995	C1/K/S	CP/LP
28.5 E	±0.1°	DFS-2	Bundespost Germany	1990	K3-4/Ka	LP
31 E	–	Turksat 1-B	Turkey	1993	K2	CP
33.5 E	±0.1°	DFS-3	Bundespost Germany	1992	K3-4/Ka	LP
36 E	–	Eutelsat I F4	Eutelsat	1988	K2/K3	LP
42 E	–	Turksat 1-A	Turkey	1993	K2	CP
53 E	*	Stat. 5/Gorizont 17	CIS	1989	C3/K5	RH
57 E	*	INTELSAT 507	INTELSAT	1983	C1/K3	CP/LP
60 E	±0.1°	INTELSAT 604	INTELSAT	1990	C2/K	CP/LP
63 E	±0.1°	INTELSAT 602	INTELSAT	1990	C2/K	CP/LP
66 E	–	INTELSAT 505	INTELSAT	1982	C1/K3	CP/LP
68 E	–	PAS-6	Alpha Lyracom	1995	C/K	–
70 E	–	Unicom F2	Unicom [U.S.]	1997	C1/K	LP
72 E	–	PAS-7	Alpha Lyracom	1995	C/K	–
77.5 E	–	Asiasat 2	Asiasat [Hong Kong]	1995	C/K	LP
78.5 E	–	Thaicom A2	SCC [Thailand]	1994	C1/K	CP/LP
91.5 E	*	INTELSAT 501	INTELSAT	1981	C1/K3	CP/LP
101.5 E	–	Thaicom A1	SCC [Thailand]	1984	C1/K	CP/LP

© 1992 MLE INC
All Rights Reserved
Post Office Box 159,
Winter Beach, FL 32971
Tel: (305) 767-4687
Fax: (305) 767-6067

World Satellite Ku-Band Satellite

U =	700 to 730 MHz	L =	1.5 to 1.65 GHz
C2 =	3.65 to 4.2 GHz	C3 =	3.65 to 3.95 GHz
K1 =	11.7 to 12. 2 GHz	K2 =	12.2 to 12.7 GHz
K5 =	11.52 to 11.56 GHz	K6 =	11.7 to 12.5 GHz
Ka =	19.1 to 20.2 GHz	CP =	Circular Polarizati...
VP =	Vertical only	HP =	Horizontal only

Figure 2.5 The allocation of the geostationary orbit for Ku-band satellites, 1992. Source: Long, 1992.

Telecommunications as material entity & spatial artefact

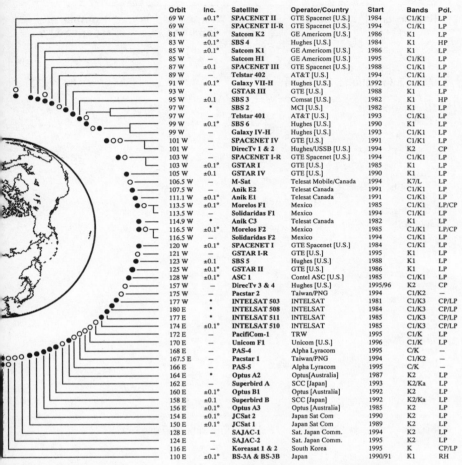

Orbit	Inc.	Satellite	Operator/Country	Start	Bands	Pol.
69 W	±0.1°	SPACENET II	GTE Spacenet [U.S.]	1984	C1/K1	LP
69 W	—	SPACENET II-R	GTE Spacenet [U.S.]	1994	C1/K1	LP
81 W	±0.1°	Satcom K2	GE Americom [U.S.]	1986	K1	LP
83 W	±0.1°	SBS 4	Hughes [U.S.]	1984	K1	HP
85 W	±0.1°	Satcom K1	GE Americom [U.S.]	1986	K1	LP
85 W	—	Satcom H1	GE Americom [U.S.]	1995	C1/K1	LP
87 W	±0.1	SPACENET III	GTE Spacenet [U.S.]	1988	C1/K1	LP
89 W	—	Telstar 402	AT&T [U.S.]	1994	C1/K1	LP
91 W	±0.1°	Galaxy VII-H	Hughes [U.S.]	1992	C1/K1	LP
93 W	*	GSTAR III	GTE [U.S.]	1988	K1	LP
95 W	±0.1	SBS 3	Comsat [U.S.]	1982	K1	HP
97 W	*	SBS 2	MCI [U.S.]	1982	K1	LP
97 W	—	Telstar 401	AT&T [U.S.]	1993	C1/K1	LP
99 W	±0.1°	SBS 6	Hughes [U.S.]	1990	K1	LP
99 W	—	Galaxy IV-H	Hughes [U.S.]	1993	C1/K1	LP
101 W	—	SPACENET IV	GTE [U.S.]	1991	C1/K1	LP
101 W	—	DirecTv 1 & 2	Hughes/USSB [U.S.]	1994	K2	CP
103 W	—	SPACENET I-R	GTE Spacenet [U.S.]	1994	C1/K1	LP
103 W	±0.1°	GSTAR I	GTE [U.S.]	1985	K1	LP
105 W	±0.1	GSTAR IV	GTE [U.S.]	1990	K1	LP
106.5 W	—	M-Sat	Telesat Mobile/Canada	1994	K7/L	LP
107.5 W	—	Anik E2	Telesat Canada	1991	C1/K1	LP
111.1 W	±0.1°	Anik E1	Telesat Canada	1991	C1/K1	LP
113.5 W	±0.1°	Morelos F1	Mexico	1985	C1/K1	LP/CP
113.5 W	—	Solidaridas F1	Mexico	1994	C1/K1	LP
114.9 W	*	Anik C3	Telesat Canada	1982	K1	LP
116.5 W	±0.1°	Morelos F2	Mexico	1985	C1/K1	LP/CP
116.5 W	—	Solidaridas F2	Mexico	1994	C1/K1	LP
120 W	±0.1°	SPACENET I	GTE Spacenet [U.S.]	1984	C1/K1	LP
121 W	—	GSTAR I-R	GTE [U.S.]	1995	K1	LP
123 W	±0.1	SBS 5	Hughes [U.S.]	1988	K1	LP
125 W	±0.1°	GSTAR II	GTE [U.S.]	1986	K1	LP
128 W	±0.1°	ASC 1	Contel ASC [U.S.]	1985	C1/K1	LP
157 W	—	DirecTv 3 & 4	Hughes [U.S.]	1995/96	K2	CP
175 W	—	Pacstar 2	Taiwan/PNG	1994	C1/K2	—
177 W	*	INTELSAT 503	INTELSAT	1981	C1/K3	CP/LP
180 E	*	INTELSAT 508	INTELSAT	1984	C1/K3	CP/LP
177 E	*	INTELSAT 511	INTELSAT	1985	C1/K3	CP/LP
174 E	±0.1°	INTELSAT 510	INTELSAT	1985	C1/K3	CP/LP
172 E	—	PacifiCom-1	TRW	1995	C1/K	LP
170 E	—	Unicom F1	Unicom [U.S.]	1996	C1/K	LP
168 E	—	PAS-4	Alpha Lyracom	1995	C/K	—
167.5 E	—	Pacstar 1	Taiwan/PNG	1994	C1/K2	—
166 E	—	PAS-5	Alpha Lyracom	1995	C/K	—
164 E	*	Optus A2	Optus[Australia]	1987	K2	LP
162 E	—	Superbird A	SCC [Japan]	1993	K2/Ka	LP
160 E	±0.1°	Optus B1	Optus [Australia]	1992	K2	LP
158 E	±0.1	Superbird B	SCC [Japan]	1992	K2/Ka	LP
156 E	±0.1°	Optus A3	Optus [Australia]	1985	K2	LP
154 E	±0.1°	JCSat 2	Japan Sat Com	1990	K2	LP
150 E	±0.1°	JCSat 1	Japan Sat Com	1989	K2	LP
128 E	—	SAJAC-1	Sat. Japan Comm.	1994	K2	LP
124 E	—	SAJAC-2	Sat. Japan Comm.	1995	K2	LP
116 E	—	Koreasat 1 & 2	South Korea	1995	K	CP/LP
110 E	±0.1°	BS-3A & BS-3B	Japan	1990/91	K1	RH

ALMANAC Wall Chart

=	2.5 to 2.6 GHz	C1 =	3.7 to 4.2 GHz
4 =	3.4 to 3.675 GHz	C5 =	4.160 to 4.198 GHz
3 =	10.95 to 11.7 GHz	K4 =	12.5 to 12.75 GHz
7 =	10.75 to 10.95 GHz	K8 =	11.7 to 11.9 GHz
H =	Right Hand Circular	LP =	Horizontal/Vertical
	Inclined Orbit		

All the world's satellites, explained, illustrated, footprinted, indexed and analysed in detail!
(Future INTELSAT deployment plans can be found on page 156 of the new third edition of the World Satellite A L M A N A C)

Table 2.5 Communications satellites, 1990

Region or organization	Number of satellites	Number of transponders
Intelsat	15	776
Other international organizations	11	36
North America		
USA	43	706
Canada	5	120
Europe	11	157
France	4	
Germany	3	
UK	2	
Luxembourg	1	
Sweden	1	
Japan	8	171
Rest of the world	48	343
USSR	33	
Australia	3	
Brazil	2	
China	3	
Hong Kong	1	
India	2	
Indonesia	2	
Mexico	2	
Total	200	2309

Sources: Wilson (1991); *World Space Industry Survey* (1991).

coincidentally with the divestiture of AT&T (Gershon, 1990). Elsewhere, INTELSAT was joined by additional international companies on a regional or organizational basis, notably EUTELSAT for European nations, INMARSAT for maritime communications, as well as the offer of services to neighbouring countries by nations such as Indonesia and Australia (Snow, 1988). Such cooperative attempts do not only reduce prices, but they relieve pressures on the restricted space and wave resources.

The demand for satellite services has saturated North America, so new launches are usually aimed at replacing existing satellites. However, demand is still growing in Europe and in Japan (*World Space Industry Survey*, 1991). Satellite networks at large have been challenged in recent years by the growing number of maritime optical fibre cables, and this issue will be addressed in the next subsection.

Maritime cables

The use of maritime cables for international, notably transoceanic, telephony has been in a historical interplay with wireless technologies. For three decades, between the 1920s and 1950s, it was poor, cumbersome and expensive radio communications that served as the sole transoceanic telecommunications technology. The completion of the first transatlantic telephone cable (TAT-1) in 1956, and the first Pacific cable (Hawaii-1) in 1957, marked a move to an accent on the more reliable, higher quality, maritime cable technology for international telephone communications. It took less than a decade for another wireless technology to challenge maritime cables, namely satellite communications, beginning service in 1965, and permitting larger bandwidth video communications, as well as faster data transmission. In that same year, however, a second transatlantic cable (TAT-4) was put into service, increasing the capacity in voice paths from 89 to 227. This cable followed the inauguration of another transpacific cable (TPC-1), which increased the capacity of international telecommunications there from 91 to 278 voice paths. However, the transoceanic transmission capacity of INTELSAT, which amounted to 150 transponders in 1965, soon dominated the market. Despite this dominance, technological improvements in wire technology permitted a steady decline in the cost of a maritime voice path, from $557,000 in TAT-1 in 1956 to just $23,000 in TAT-7 in 1983, bringing the total available transatlantic maritime voice paths in 1983 to over 18,000. The total maritime voice paths across the Pacific at that time was much lower, reaching about 3,600 (Staple, 1991).

The late 1980s and 1990s have been marked by three important changes. First, there emerged a renewed competition between wireless (satellite) technology, and cable technology, this time in the form of fibre-optics. Second, an increased demand for transpacific communications has evolved, reflecting the growing importance of the Pacific Rim in the global economy. Third, competition was introduced in both satellites and maritime cable communications.

In 1983, 189 sea cables were in use world-wide, of which only 15 per cent were intercontinental. The highest maritime cable density was in Europe with 37 per cent of the world's cables. In 1989, 19 fibre-optic maritime cables were in use or planned (Hottes, 1992). The uneven geographical distribution of sea cables at large, and that of fibre-optic cables in particular is evident in Figures 2.6 and 2.7. One may identify three cable-rich world regions, namely the Atlantic, the Pacific, and the Mediterranean. Only few connections exist between South America and Africa, and none from these two areas to the Pacific Rim. The cable maps represent therefore, the more general web of connections among the three world cores (North America, Europe, and the Pacific Rim), and the connections of the rest of the world with one or two of them.

Fibre-optics versus satellites: Fibre-optic cables usually extend from one heavy telecommunications-consuming world city to another, so that networks from other domestic locations leading to such a gateway may consist of slower and restricted copper or coaxial wires and cables. Satellites, on the other hand, connect multiple locations equally with the use of dish antennas (Moss, 1987a). Fibre-optic cables permit speedier and wideband communications, immuned from electromagnetic interference, and the cost per maritime voice path is expected to go down to $2,000 by 1993 (Staple, 1991; Gershon, 1990; Hottes, 1992). Thus, investments in cable systems now largely exceed those made in satellites, a trend opposite to that which prevailed during the last ten years. It resulted in a decline in INTELSAT revenues as of 1989 (*World Space Industry Survey*, 1991).

The tremendous growth in both current and projected availability of transoceanic channels, and the emerging competition between satellites and cables are presented in Figure 2.8. It is

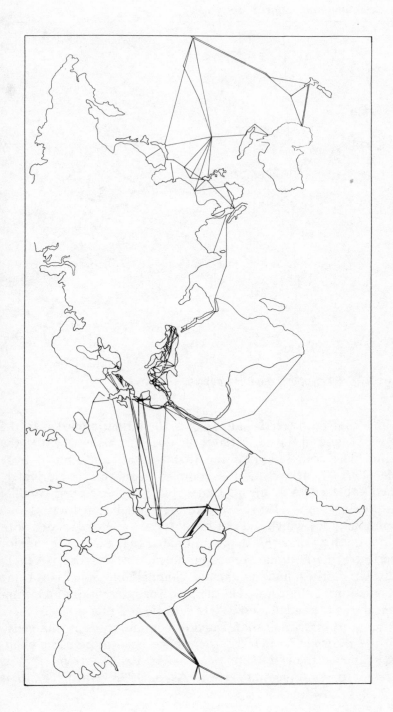

Figure 2.6 Existing and planned maritime cables, 1988.

50 Telecommunications & geography

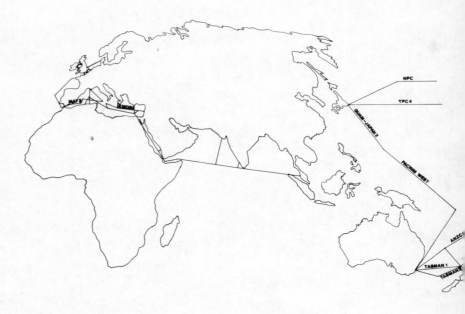

Figure 2.7 Fibre-optics maritime cables, 1992.

obvious that both trends started over the Atlantic market in the late 1980s, and in the early 1990s they became also present in the Pacific. The completion of the first transatlantic optical fibre cable (TAT-8), as well as the completion of the first privately owned cable (PTAT) meant that cable paths exceeded satellite transponders in 1989. However, satellite transatlantic communications which did not grow for a while, responded with rapid growth as of 1990. It was estimated that each of the major American international telecommunications carriers (AT&T; MCI; US Sprint) had by then a channelling capacity for the transmission of the whole US demand for international telecommunications (Langdale, 1989a). It is projected that by 1996 there will be more maritime than satellite communications channels.

In the transpacific market, there have been more cable channels available than satellite ones as of 1990, a trend that is expected to continue until 1994. As of 1996, when TPC-6 is

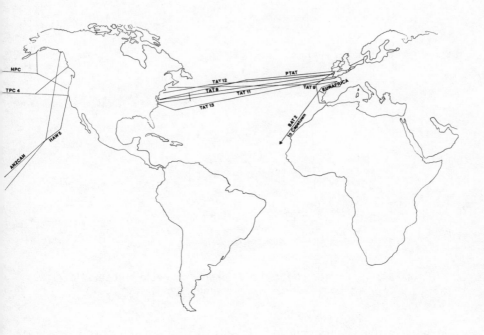

scheduled for completion, with a capacity of 605,000 voice paths, satellite communications will become secondary to cables.

Satellite communications presents several advantages (Langdale, 1989b; *World Space Industry Survey*, 1991). It is more cost effective for long-distance low volume routes, and is thus vital for Third World countries. It is also very useful for television broadcasting and mobile telephony. On the other hand, fibre-optics offer more security in the transmission process; they are not dependent on launching failures and on an international allocating agency as satellites are. These and other qualities make cables especially attractive for point-to-point leased networks. In the long run, one may view satellite and cable networks as complementary to each other, and as providing back-ups for one another.

The Atlantic versus the Pacific: Figure 2.8 shows high growth

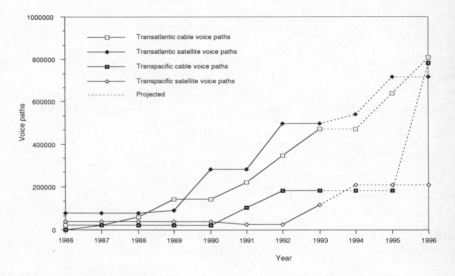

Figure 2.8 Estimated capacity of transoceanic cable and satellite systems 1986-96. Data source: Staple, 1991.

rates in the Pacific market. It was estimated that for the period between 1982–96, 51.3 per cent of all planned investments in maritime optical fibre cables will be made in the Pacific (*World Space Industry Survey*, 1991). By 1996 it is expected that the number of cable circuits in the Pacific will almost equal that across the Atlantic. However, given the larger satellite communications availability over the Atlantic, the latter will retain its seniority in transoceanic networking. The growing communications capacity in the Pacific Rim demonstrates the increasing importance of the region in the global economy, as well as the growing immigration from Asian-Pacific nations to the US.

Competition: The late 1980s marked the introduction of competition into the cable market, a trend which we noted also for satellite communications (Langdale, 1989b; *World Space Industry Survey*, 1991). Early cables were jointly sponsored by PTTs and AT&T. The Private Transatlantic Telecommunications (PTAT) fibre-optic cable was jointly sponsored by US

Sprint (US) and Cable & Wireless (UK), connecting New York and London as of 1989. In 1991 the North Pacific Cable (NPC) was put into service, jointly owned by Cable & Wireless (UK) and International Digital Communications (Japan). These new, as well as veteran, investments in international telecommunications networks will continue to push communications costs down, a trend which, for its part, has resulted in increasing demands.

A special dimension of networking is leased networks, studied by Langdale (1989a). These are usually point-to-point circuits leased by transnational corporations from common carriers for a flat fee, in order to provide permanent connections among computers in different locations. The topologies of leased networks are similar to those of switched networks. A centralized network, centred on global headquarters, will usually have a star form, whereas a global network would adopt the mesh form. The creation of regional hub-and-spoke networks may take the tree form. The choice among network patterns depends on company needs, so that communications to just a handful of countries does not require regional nodes, while a more centralized pattern may promote economies of scale.

Nodes

Nodes of telecommunications systems may consist of telephone exchanges, microwave relay towers, or places which house concentrations of a variety of telecommunications equipment. Telephone exchanges based on analogue-mechanical equipment used to occupy large buildings, so that telecommunications nodes in urban settings had a noticeable locational and real estate dimension. The introduction of digital equipment has meant, among other things, a miniaturization and full automation of telephone exchanges, so that they require small areas, very often in basements or other peripheral areas of construction complexes.

Microwave relay stations and antennas may be located on roofs of high-rise office towers, hotels, and even residential buildings, and as such they have become an integral, though not the most beautiful part, of contemporary city landscapes. Since microwave relay stations have to 'see' each other in order to transmit signals, their geographical effectiveness increases the higher their locational altitudes become. Elevated mountain peaks are, therefore, desirable sites, and land-use conflicts may evolve between the telecommunications industry and other sectors who may make use of elevated areas. The latter constitute mainly military communications systems and outposts, and touristic observation and recreation areas. Such conflicts may emerge especially in small countries, as is the case in Israel.

The most important telecommunications nodes in contemporary technological settings are, by far, concentrations of telecommunications equipment at various geographical scales. These are *smart buildings* at the micro, 'point' scale, *teleports* at the meso, urban scale, and *telecottages*, at the macro, regional scale. All three types of nodes serve as *hosting locations* (Gottmann, quoted by Abler, 1991, p.34) for advanced services, and as *gateways* (Hepworth, 1990, p.197) for the distribution of information, notably to and from international points.

Smart buildings

Smart buildings are new office buildings equipped with built-in sophisticated telecommunications systems, or older upgraded office buildings mainly in downtown areas. Such buildings provide dedicated plug-in fast telephone lines for computer data, or ISDN services, in addition to traditional telecommunications services. They may further provide automatic answering services, videoconferencing facilities, as well as automatic energy-saving controls (see also Moss, 1986a; 1986b, pp.393–94). Such buildings are normally equipped with their own dish-antennas for satellite connections, and/or fibre-optic cables leading to long-distance fibre-optic cable systems. They further offer special inter-office communications systems, including express courier pick-up services.

Smart buildings demonstrate the interdependencies among telecommunications, information technology and the service economy, since they serve mainly producer services of various types, such as computer companies, telecommunications companies, consulting firms, as well as financial, marketing, accounting, and legal services. This interdependence is presented in smart buildings at its tightest geographical level, namely all elements are concentrated in one building.

Teleports

Teleports constitute larger concentrations of telecommunications equipment, serving an adjacent service or industrial park built in conjunction with the teleport, or it may serve other producer services or industrial facilities located at the community at large (Phillips, 1986, p.84; Hanneman, 1986). As such, the 'port' itself is fully automatic, and is, therefore, different from transportation terminals for land, air or maritime travel, which are typified by their congestion of people and cargo. On the other hand, teleports are similar to transportation ports in their geographical service as gateways for information and in their service as hosting locations for various economic activities attracted to such gateways. Furthermore, teleports, like transportation ports, may serve a variety of users (see Hanneman, 1986, p.6). Thus, in Amsterdam and Rotterdam teleports are meant to enhance the Dutch seaport industry; in New York it initially served the financial community; in San Antonio, Texas, it was supposed to attract high-tech industries; and in London (Docklands) and Paris (Ile-de-France) teleports were part of more extensive efforts to develop regional hubs (Hepworth, 1990, p.197; Moss, 1991).

Teleports are especially attractive for heavy users of international telecommunications services, since location near them may permit direct access to the international network, thus eliminating domestic lags in communications networks. They are further based on the introduction of competition into international telecommunications (Bakis, 1988).

The teleport concept was initially developed by the Port Authority of New York and New Jersey in the late 1970s, when a telecommunications park was built in New Jersey connecting various Manhattan office buildings via fibre-optic links to any

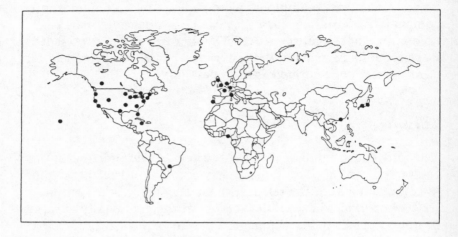

Figure 2.9 Major teleports, 1986. Source: Warf, 1989.

requested point. In 1991 this teleport operated 20 earth stations (with a capacity for 30). The fibre-optic network was 174 miles long, serving 247 customers in 162 office buildings (Moss, 1991). New York was a 'natural' location for the first teleport, which served a dispersed clientele, the city being the largest and most sophisticated concentration of the service economy and also because of the congestion of the local telephone system.

In the late 1980s more teleports existed in the US than in any other continent or country (Figure 2.9). This may be attributed to the nature of supply and demand in the US. The US has developed its contemporary long-distance telecommunications system on a competitive-business basis since 1982, so that offering sophisticated services has typified the supply side of the system. From the perspective of demand, the US has been the first and fastest country to develop a service economy (Kellerman, 1985). Thus, an early need for advanced telecommunications services emerged in major metropolitan areas, before the city-wide telecommunications system could offer these services universally. Not meeting such demands could have meant a decline in the ability of firms to compete in domestic business.

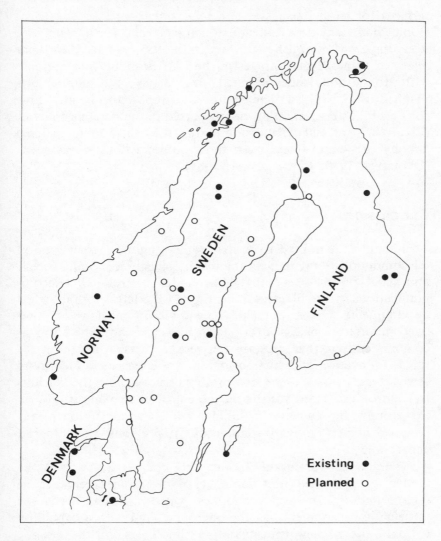

Figure 2.10 Telecottages in Scandinavia, 1988. Source: Qvortrup, 1990.

Telecottages

The Scandinavian *telecottages* (formally called Community Teleservice Centres) constitute a similar idea to teleports for rural areas. These are centres which offer advanced telecommunications services for rural businesses and households, such as fax, electronic mail, and data transmission (Qvortrup, 1990). The first telecottage was established in Sweden in 1985, and by 1988 there operated some 25 telecottages in the four Scandinavian nations, with 20 additional ones planned for 1988/89, and another one planned for Sri Lanka. The centres are supposed to be attractive for new industries, and they are dispersed in the rural peripheries (Figure 2.10). Centre ownership is shared by local municipalities and by private business, sometimes subsidized by government (Qvortrup, 1990; Parker *et al.*, 1989).

Conclusion

This chapter attempted to present wide-ranging perspectives on telecommunications, namely the technological, economic, social, and geographical ones, moving from the inception of telecommunications in the nineteenth century to the latest technological developments. From a technological viewpoint it seems that telecommunications, as information mover, is now the form of communications that receives the most extensive attention, as far as innovation and transformation are concerned. This trend seems logical, given the maturation reached by the various transportation means for the movement of people and cargo, on the ground, by sea, and in the air. These transportation means received higher innovative attention in previous technological generations.

From an economic standpoint, this chapter reaffirmed the tendencies presented in Chapter 1, namely the emergence of service and information economies, with particular needs for information movement, on the one hand, and with a capability to make use of new transmission technologies, on the other.

Telecommunications has several social characteristics: it is a device which may be equally accessed by all household members; it is also a 'flattening agent' for space and time notions; and it is further a facilitating factor for a more dispersed distribution of extended families.

Geographically, we noted the increased importance of the Pacific Rim; the locational separation between production and controlling; the increased levels of travel; and the growing importance of controlling cores.

Telecommunications may thus be seen as a technology which is simultaneously undergoing major changes as well as an agent contributing to change in the various contexts within which it operates. This double character of current developments in telecommunications will continue to accompany us in the following chapters.

Chapter 3
Spatial Dynamics of Telecommunications

The 'spatial dynamics' of telecommunications includes three basic dimensions, namely flow, diffusion, and movement. Flows refer to flows of information through telecommunications channels. Diffusion relates to the adoption of telecommunications innovations along time and space, and movement is interpreted here as the possible role of telecommunications as a substitute for the movement of people. These three dimensions will be discussed in the following sections.

Flows

'Society can be viewed as a series of nested message flows. Any group of people who communicate regularly generates a matrix of information flows' (Abler and Falk, 1981, p.13). Also, 'telecommunications links provide a potential for communication. Nodes provide access to that potential. Flows of messages among places reflect the degree to which that potential is realized, and fluctuations reveal how communications patterns among places change from time to time' (Abler, 1991, p.35). Flows of information are, therefore, of crucial importance for a geography of telecommunications, and their study constitutes a direct continuation to an understanding of transmission media, networks, and nodes. However, little empirical attention has been directed to flows of information, given the dearth of data provided by PTTs and telephone companies on the number, length, origins and destinations of telephone calls.

Conceptual developments concerning flows reflect these circumstances, so that little work has been done so far, and most of it refers to, or has been developed for, international telecommunications, for which some aggregate data are available. The discussion of information flows will treat these flows as a four-phase process. First, information types will be outlined, followed by an exposition of the various characteristics of information flows. Later, the possible barriers for information flows will be highlighted and classified by pattern and type. Finally, several patterns which may be induced by information flows will be outlined (Figure 3.1).

Information types

It is possible to classify information into four major groups by contents of message (Kellerman and Cohen, 1992):

Economic information: These are messages which relate to the transfer of goods, services and people as tourists, as well as to the transmission of funds.
Institutional information: These are calls which are generated by governments, non-profit organizations, cultural and academic institutions.
Social information: Exchanges among people sharing family and social ties.
Produced information: This type relates to information as a commercial product *per se*, such as computer programs, raw or processed data, radio and TV reports and programmes.

This classification implies an urban classification of geographical origins and destinations of flows. Thus, cities in the upper levels of the urban hierarchy would generate more economic and institutional information, and world cities would, in addition, generate and receive more produced information. By the same token, residential areas may produce more social flows of information, whereas CBDs would be typified by economic and institutional information flows. The geographical origins of business calls are, therefore, more concentrated in

62 Telecommunications & geography

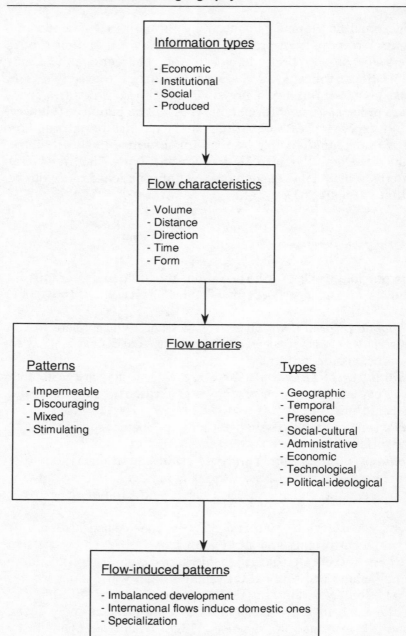

Figure 3.1 Dimensions of information flows through telecommunications.

downtowns and major suburban office and manufacturing parks, whereas the origins of residential calls may be as widely distributed as population is. The role of residential areas in the flow of information has increased profoundly. It was estimated that in the early 1960s, 20 per cent of US international calls originated in households, and 80 per cent of calls were business or institutional (Lerner, 1968). By the late 1980s it was estimated by AT&T that about two-thirds of American outgoing international phone calls were placed from households. Moreover, residential international calls made from Israel were found to be about one-third longer than international business calls (*Bezek*, 1991). Thus, residential areas may produce more and longer international calls than business areas. The growing share of the residential sector in the international flows of information is related to the introduction of a reliable, direct-dialled, and cheaply priced service.

The four types of information are not as clearly separated in the daily conduct of telecommunications. Businesses might be operated by family members, businessmen may maintain social relations as well, and governments deal with exports and imports. The first three types of information are strongly related to the movements of commodities, capital and people, while only the last one is a 'pure' transfer of information. Though no separate accounts exist for the four types of information, the transfer of produced information over the public telephone system is smaller than the other three types, at least in developed countries. It was estimated that the use of telephone lines worldwide is divided between 90 per cent for telephone calls and only 10 per cent for data-transfer (*The Economist*, 1990).

All four types of information may use the same infrastructure of microwaves, satellites and cables, with different or similar transmission media or terminals. However, as mentioned earlier there are different network transmission preferences for video and data, relating to speed, bandwidth and reliability, whereas telephone calls are the most flexible form of information flow. These differences may result in separate prices for the different uses.

Flow characteristics

Information flows can be typified by a variety of characteristics, namely: volume, distance, direction, time, and form.

Volume: Volume was singled out as the most basic dimension for the study of communication flows at large (Deutsch, 1956, p.145). Needless to say volume also refers to growth in information flows. Information flows have experienced constant growth world-wide, going hand in hand with the tremendous recent developments in telecommunications infrastructure, as well as with the more varied uses offered by telecommunications systems. In the US, domestic calls increased by about 5 per cent in 1987, though the number of lines grew by around 2.5 per cent only, or in other words, flows grew twice as much as subscribers. In Israel, total calling units went up by 11.8 per cent in 1989, whereas the number of lines grew by 4.2 per cent only, so that information traffic increased almost three times more than connected stations (*Bezek*, 1991). Such figures attest to both growing demands and expanding supplies, often coupled with declining prices.

Distance: It seems obvious that a distance-decay pattern would apply to the relationship between the volume of calls, on the one hand, and distances from call-origins to call-destinations, on the other. Such a relationship was found in a Dutch study employing data on equally-priced interregional phone calls (Rietveld and Janssen, 1990). It was also found that 75 per cent of all calls placed in Manhattan were to points within Manhattan, whereas in other New York boroughs the respective percentage was 60 (Moss, 1986b, p.390). This obvious relationship holds, however, to volumes only, but not necessarily to growth rates. Thus, in the US, 88 per cent of the domestic calls placed in 1987 were local, but they grew by 2 per cent only from 1986, whereas the remaining 12 per cent of long-distance calls represented a growth rate of 26 per cent (*The Economist*, 1990). The share of toll-calls to local ones in the US has been on the rise since the 1920s (Figure 3.2). Thus, for 1900–20 the ratio between the number of local calls to long-distance ones was 1:40, falling to just 1:9 in 1987. The share of international

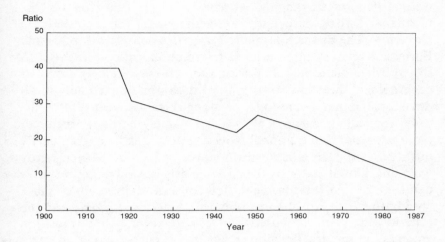

Figure 3.2 The ratio of toll to local calls in the US (toll = 1), 1900–87.
Data sources: Abler, 1977, p.335; *The Economist*, 1990.

transmissions was high in fax communications, where it amounted to 15 per cent of all pages transmitted in the US in 1990 (Martin, 1991). US international calls (excluding Canada, Mexico, and the Caribbean) grew at a rate similar to long-distance ones, from 1986 to 1987 (Kellerman, 1992b). By the same token, domestic calls in Israel grew by 7.3 per cent from 1989 to 1990, compared with 22.8 per cent for international calls for the same year (*Bezek*, 1991).

Higher growth rates for long-distance calls in the US reflect the growing competition among long-distance telephone companies which yielded lower prices. This factor applies also to the international arena, where improved technologies permitted price reductions even in non-competitive markets. The growing global economy has brought about noticeable high growth rates in international calls. In Japan, the growth rate for April 1990 to March 1991 was 23 per cent, and growth rates of over 20 per cent were experienced in 1990 by other Pacific countries, such as Singapore, Taiwan, Hong Kong, and Thailand. On the other hand, the growth rates for major European countries for 1990 were more modest, ranging between 12–15 per cent (for the UK, Germany, Switzerland, and The Netherlands) (Staple, 1991).

Direction: The distinction made between Asian growth rates and European ones in international telecommunications leads us to the third dimension of information flows, namely that of geographical direction. Information does not flow equally in all directions from a given node, or along certain routes. This is the case for all geographical scales, namely urban, regional, national, or international ones, reflecting geographically differentiated economic, institutional or social ties. The currently emerging global patterns are particularly striking. Seventy-five per cent of the international flow of information takes place among the three world cores in North America, Europe, and the Pacific Rim (*The Economist*, 1991). However, some transition emerges in the traffic among the three cores, along the two major routes, namely the Pacific and the Atlantic. In 1990, 59 per cent of the traffic in the INTELSAT satellite system came from the Atlantic region, 23 per cent was generated by Indian Ocean countries, and 18 per cent came from the Pacific. The average annual growth rates for 1984–90, showed a growing tilt towards the Pacific. The growth rate for the Pacific was 13 per cent, for the Indian Ocean it was 7.9 per cent, and for the Atlantic it was 7.5 per cent (*World Space Industry Survey*, 1991). This trend reflects once again the changing geographical emphasis of the global economy towards the Pacific.

Time: The temporal dimension of information flows refers to two aspects: one is the distribution of calls by times of calling, and the other is the length of calls. In many countries the day is divided into two or three rating periods, so that calling during business hours is more expensive than calls placed at night. The same principle also applies to a distinction between weekdays and weekends. The reason for such differential rates is the attempt by phone companies to distribute calls more evenly, so that more efficient and less congested use will be made of the telephone infrastructure by diverting social calls to non-business hours and days.

The length of calls may vary geographically, or more precisely, by distance of calling. The Israeli common carrier reported for 1990 that average local calls were 2.6 minutes long,

Spatial dynamics of telecommunications 67

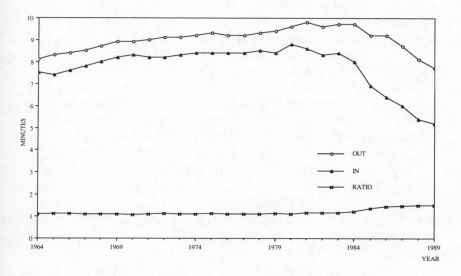

Figure 3.3 The average lengths of incoming and outgoing US international phone calls, 1964–89. Source: Kellerman, 1992.

average regional ones lasted 2.8 minutes, interregional calls had an average length of 2.9 minutes, and international calls were the longest, averaging at 3.8 minutes (*Bezek*, 1991). An average international call was about time-and-a-half longer than an average domestic one. Furthermore, variations in the pricing of calling-time by distance create big differences between the geographical distribution of calls and the revenues they generate. Thus, 63 per cent of the calling minutes in Israel in 1990 were local calls, but they amounted to only 33 per cent of the charging units, whereas international calls amounted to only 1.2 per cent of calling minutes, generating 25 per cent of the charging units! (*Bezek*, 1991).

It is of interest in this regard to compare trends in the average length of incoming and outgoing US international calls (Kellerman, 1992a). These trends for 1964–89 are presented in Figure 3.3. One may notice a gradual increase in the average length of outgoing calls from 8.3 minutes in 1964 to a peak of 9.8 minutes in 1981. By the same token, the mean for incoming call length increased from 7.5 minutes in 1964 to 8.8 minutes in 1980. Both categories declined in the 1980s, so that the mean outgoing call length was only 7.7 minutes in 1984, and 5.2 minutes respectively for incoming calls. Both values are much lower than those for 1964. The ratio between the two measures changed drastically. Between 1964 and 1981 an average outgoing call from the US was 10 per cent longer than an incoming one. Since then, the average length of calls originating in the US increased considerably and quickly, so that as of 1987 it has been 50 per cent longer than the mean incoming message!

These trends may be explained by a mixture of market forces, social habits and technology. American call rates have declined drastically since the introduction of competition into the international calling market, in the mid-1980s. Americans have incorporated the telephone into social and business life earlier and more extensively than other nations, so leading to longer calls. The introduction of extensive fax services has probably made incoming business calls shorter, by using fax rather than longer, and therefore, more expensive, voice phone calls.

Form: The last characteristic of information flows is its form. It has been noted earlier that about 90 per cent of world-wide traffic constitutes spoken phone calls, and the rest is data transmission. However, spoken calls have increased at an annual rate of 7 per cent, whereas data transmission has grown by 25–30 per cent annually (*The Economist*, 1990). These trends, as well as those mentioned earlier, imply that the image of the telephone service as a device geared mainly for domestic spoken information flows, is about to change.

Flow barriers

'Telecommunications is a particular type of infrastructure that serves to reduce the negative impacts of geographical barriers' (Nijkamp *et al.*, 1990). Still, information flows via telecommunications may be blocked, slowed down, speeded up, concentrated or diluted by various domestic or international barriers. These barriers may be cross-classified by pattern or type.

Nijkamp *et al.* (1990) proposed four spatial patterns of barriers for information flows:

1. *Impermeable barriers*, typically due to political factors, prohibit any transborder flow. (Broadcasting may be an exception, as it can overcome most borders).
2. *Discouraging barriers*, reflecting a discontinuous increase in communication costs.
   2.a. *Symmetric discouraging barriers*: equal in both directions.
   2.b. *Asymmetric discouraging barriers*: not equal in both directions.
3. *Mixed barriers*, reflecting a discontinuous increase in communication costs in one direction, and a decrease in the other direction.
4. *Stimulating barriers*, reflecting a discontinuous decrease in communication costs.
   4.a. *Symmetric stimulating barriers*: equal in both directions.
   4.b. *Assymetric stimulating barriers*: not equal in both directions.

These patterns may apply to any of the following eight types of barriers: geographic, temporal, presence, social-cultural, administrative, economic, technological and political-ideological.

Geographical barriers: Though telecommunications by its nature reduces the impact of 'the tyranny of space', it might still happen that maritime or topographic circumstances would avoid the construction of a full or partial telecommunications infrastructure. This could happen especially when demands on both ends of a connection do not justify it, so that other, well-connected regions would enjoy an advantage.

Temporal barriers: Time may serve as a barrier in intercontinental communications, if common business or leisure hours between two continents are scarce. This barrier has brought about shift work in financial services, when daily participation in stock market activities in the three global centres of New York, London, and Tokyo is sought. On the other hand, the mass introduction of fax service and e-mail has permitted speedy exchanges of messages over the telephone system without a physical co-presence of the communicating persons.

Presence barrier: The presence barrier can be at least partially solved through the videophone. However, as it has been argued in the previous chapter, it is not yet clear whether socially communicating people would actually like to constantly remove this barrier. Even in the use of videoconferencing for remotely conducted business meetings, the very lack of physical presence of the participants and the non-existence of the accompanying gestures and social atmosphere it implies, amount to a still remaining presence barrier.

Socio-cultural barriers: Social-cultural barriers find their expression mainly in the form of language disparities. Such disparities may now be overcome, to some degree, by the use of fax, rather than voice telephone, so that written messages can be translated. It was estimated that about 50 per cent of the international telecommunications traffic between the US and Japan is channelled via fax. This was attributed to the language barrier, as well as to the long time interval between the two countries (Langdale, 1989b). It was further shown that the Germans prefer to converse with the German-speaking Austrians, rather than with the French and the Italians who have ranked Germany as their most frequently called country (Kellerman, 1990). The language barrier can find its expression not only in international telecommunications but in domestic calls as well, especially in multi-language countries. It was shown that interregional calls in Belgium were effected by the linguistic barrier between Flemings and Walloons (Klaassen *et al.*, 1972), and similar findings were shown for Switzerland (Rossera, 1990).

Administrative barriers: The flow of information may be influenced by adminstrative involvement, mainly through what is frequently termed as regulation. Abler (1991) claimed that 'regulation exerts more control over what services are available to what classes of customers at what places at what costs than any other single variable' (p.38). This is so mainly in the US, where the nationwide Bell telephone system has been divided into seven Regional Bell Operating Companies (RBOCs, or the 'baby-Bells') in 1982. A well-known geographical dimension of

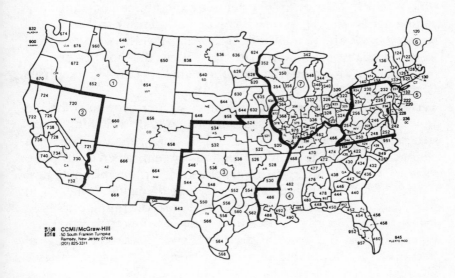

Figure 3.4 US Local Access and Transport Areas (LATAs), 1984. Source: Abler, 1991, p.39.

this process are the American Local Access and Transport Areas (LATAs) (Abler, 1991) (Figure 3.4). The 187 LATAs define areas of monopolistic service for the local companies operated by the RBOCs, whereas flows between LATAs are the responsibility of competing long-distance carriers.

Economic barriers: Economic barriers refer mainly to the impact of rates on information flows. In most countries calling rates in general, or at least local calling prices, are regulated, so that economic barriers go mostly hand in hand with regulation or administrative barriers. Rates may be scheduled by distance, by length or by both measures, as is the case in many countries. Telephone companies may sometimes use policies of geographical cross-subsidization, so that rates for calls between major cities become higher in order to keep prices down in less frequently used links (Langdale, 1983). Such policies have been prohibited by various regulating agencies.

Special contemporary problems are the bi-national agreements administered by the International Telecommunications Union

(ITU), which govern the payments by the phone company of a calling party to the phone company of the subscriber in the country of destination. These agreements, which were designed when competition for international calling did not yet exist in any country, may restrict price declines in this market (see Chapter 7).

Technological barriers: Technological barriers for information flows may consist of shortages in network or switching infrastructure or a lack of direct-dialling. As such, these barriers are endogenous to the telecommunications system, whereas all other barriers are exogenous. Currently the most striking technological barrier in developed countries is the lack of digital switching in many areas. In the near future, the lack of fibre-optics and ISDN technology might become a serious technological barrier, especially for the transmission of data. These technological barriers exert noticeable impacts on regional development (see Chapter 5).

Political-ideological barriers: Political-ideological barriers may prevent the free flow of information among nations. Pool (1990, p.13) warned that 'governments that try to stop the international electronic flow of information by mercantilist policies will find that they pay a considerable price in productivity for doing so; they will lose out to competing countries that allow free use of any information'. It was found that calls between European countries are only 30–40 per cent of calls made to the same distances within countries (Rietveld and Rossera, 1992). Sometimes, governments may attempt to impose an impermeable barrier for information flows, as has been the case in the Israeli–Arab conflict. However, the computerization of international telecommunications, permits to bypass and penetrate this barrier, by using third-party computerized switching services for direct-dialling.

Flow-induced patterns

There are several geographical patterns which together with other factors might be directly attributed to information flows and to barriers for such flows. The most notable ones are:

imbalanced interregional or international development, domestic flows being induced by international ones, and a local specialization in information production. These patterns will be highlighted within their proper geographical contexts in Part 2. The three flow-induced patterns have been shown to be interrelated, at least at the international level (Pool, 1990, pp.111–48). Imbalanced economic development among nations may serve as both factor and outcome of information flows, which are dominated by leading countries in the global economy. However, enhanced uneven international flows of information involve 'a cycle in which the proportion of international flows grows first, and then domestic flows catch up' (Pool, 1990, p.137). This applies even to 'telecomputing', so that initial dependence on the computing powers of foreign countries may accelerate the development of computing in the dependent country.

Growth in flows from any country or city may, eventually, accentuate a specialization in information production. Such specializations may be influenced by comparative advantage and complementarity, similarly to the rationale for international trade in commodities (Kellerman and Cohen, 1992). Pool (1990, p.147) apparently thought that specialization in information production and flow may be directed, through the following prescribed rules: choosing an activity which may lead the country to a centre of that activity; cheap importing of the knowledge required for the development of professionals; and the removal of any exogenous and endogenous restrictions on information flows as well as travel.

The diffusion and penetration of telecommunications innovations

In reviewing the diffusion and penetration of telecommunications innovations we will first take a look at this diffusion process from a geographical perspective, followed by a presentation of comparative trends for major technologies, namely the telephone, fax, personal computers, and cable television.

The diffusion process

From a geographical perspective, the diffusion process for telecommunications technologies consists of three phases, namely preconditions, diffusion patterns and results.

Preconditions: Preconditions for a widespread acceptance of an innovation, as far as users are concerned, are three. A new innovation 'must be cost-competitive with other ways of doing things; it must be compatible with users' skills as well as their work or home environment; and, it must provide a specific service concept which the user values' (Carey and Moss, 1985). These requirements may apply to individual users everywhere, and another one may usually relate more to large cities, namely overcoming of the so-called *utility-penetration paradox* (Abler, 1991, p.37). In other words, a new technology must surpass a certain threshold before it achieves self-sustained growth. Enough people at enough places have to be connected in order to convince themselves as well as others that it is worthwhile to buy the new service/technology, and this process requires central direction. Both the pioneering customers, the potential future imitating customers and central direction are normally located in major cities.

Diffusion patterns: Diffusion processes may take spatial forms of either a contagious pattern, so that adopters present some spatial continuity along time, or a hierarchical one, so that adopters spread along the urban hierarchy through time (see e.g. Abler, Adams and Gould, 1971). The diffusion of telecommunications innovations has been proven once and again to follow the hierarchical pattern. In the more general sense this was argued by Brooker-Gross (1980), and by Abler (1991). The hierarchical pattern of diffusion obviously represents the higher demand for services in and among larger cities. It further attests to the basic nature of telecommunications as means for the connection of places, so that infrastructure for new technologies connects first large cities located apart from each other. Hierarchical patterns evidently emerged at the intraurban-residential scale, as well. Thus, the social distribution of the telephone was class-specific, preferring nucleations of wealthier people, rather than connecting all residences crossed by telephone lines (Martin, 1991). Specifically, the hierarchical pattern was shown to be the case for the diffusion of various technologies in the US, such as the telephone (Langdale, 1978; 1983), commercial television stations (Berry, 1972), cable television (Brooker-Gross, 1980), contemporary competitors with AT&T long-distance services (Langdale, 1983), AT&T digital data service (Langdale, 1983), and the inter-university BITNET e-mail system (Kellerman, 1986a; 1986b).

Geographical results: The geographical results of the hierarchical pattern of the diffusion of telecommunications innovations involve a process of circular and cumulative growth in the large cities (see Pred, 1977). Such cities serve as innovation incubators and as places for the overcoming and controlling of the utility-penetration paradox. When an innovation eventually diffuses to smaller places, new innovations incubate already in the larger cities (see also Langdale, 1983).

Comparative diffusion trends

Table 3.1 presents data on current penetration rates, as well as on the duration of the availability of major residential telecommunications technologies in the US. American society is of some interest in this regard, since all these technologies were first introduced in the US. Television, as a one-way mass medium has the deepest, almost universal, penetration, though it is much 'younger' than the 'older', personal, two-way telephone, which is second to television in its penetration rate. Another difference between the two appliances is that the first does not require any use charges in the US, neither in form of annual licensing fees, nor in form of charges per usage. This seems also to be the reason for the fast diffusion of the new VCRs, when compared to cable TV, since the latter involves monthly payments. From another perspective, VCRs do not constitute a stand-alone technology, as they deepen the use of the standard TV, and as they do not require any special skills for their operation. Home computers present an opposite case despite their varied uses.

Table 3.1 US household penetration of major telecommunications technologies 1990

Technology	Years available	Penetration (%)
Telephone	114	93.3
Television	63	98.2
Cable television	39	55.7
Video cassette recorder	15	67.5
Personal computer	14	15.0

Data sources: US Bureau of the Census (1991); Beniger (1986); Carey and Moss (1985); McCarroll (1991).

In the UK, similar, almost universal penetration rates for television sets were reported for 1984 (98 per cent), and much lower ones for telephone services (80 per cent). The penetration rates for home computers (23 per cent) were much higher than in the US, and were considered the highest in the world (Batty, 1988). The considerable difference between the two countries in this regard may be explained by the availability of the videotex Prestel service in the UK, and maybe by a deeper penetration of personal computers to work places and schools in the US, so that many customers may not wish to have them at home as well.

The telephone

The diffusion of the telephone has not been equal among developed nations even in recent decades. In 1956, the US still had half of the world's telephones, though in 1950 only half of American households enjoyed a telephone service (Phillips, 1991). In 1960, just half of the skilled workers in West Germany had one, and in 1975 there were only six million telephones in France (Daly, 1991, p.40). The US has constituted a preferred case for the study of telephone diffusion, because of the relatively fast pace of geographical diffusion of the telephone there despite the wide extent of the country, and because of the private ownership of telephone services in the US. Thus, in 1876, following the introduction of the telephone,

3,000 telephones were installed, growing to one million telephones in 1899, just 23 years after its introduction (Marvin, 1988, p.64).

From a sectoral perspective, the telephone diffused primarily in the business community during its first 50 years of growth (Carey and Moss, 1985). Geographically, the first commercial inter-city link opened in 1884 between New York and Boston, followed within five years by lines from New York to Albany and Buffalo, to Washington, DC, and to Philadelphia. Between 1890–94 the system extended westward to Cleveland and Chicago. It was only in 1915 that a transcontinental service from New York to San Francisco became available, following the completion of the separate eastern and western systems (Langdale, 1978; Abler, 1977). Further diffusion of the AT&T system along the urban hierarchy was accompanied by filling geographical gaps (Figure 3.5). Later, in the twentieth century, government subsidization permitted a universal diffusion of telephone service to rural areas (Brooker-Gross, 1980).

From a technological standpoint, the telephone was able to win over the telegraph because of the immediacy and privacy it could offer, in addition to its wide geographical spread compared to the more limited inter-city telegraph service (Langdale, 1978). It was, however, for the telegraph to provide the first nationally integrated communications system, in form of the news wire services in the early twentieth century (Brooker-Gross, 1981). Technological improvements from 1894, when the Bell patent expired, until 1907 were encouraged by competition, whereas later efficiency has become the major technological motive (Langdale, 1978). The new round of competition in the long-distance market which opened in the mid-1980s, was once again typified by rapid technological improvements.

The fax (facsimile)

The fax was invented in 1843, much earlier than the telephone (1876), and it was first based on telegraph transmission (Table 2.1). In 1902 it became possible to transmit pictures and maps, not just documents (Simpson, 1984). The telegraph technology

78 Telecommunications & geography

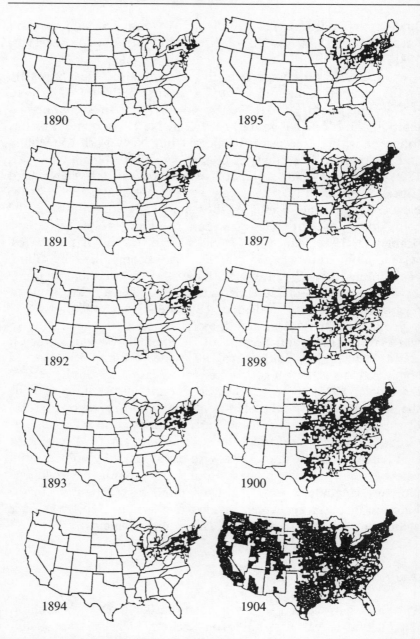

Figure 3.5 Growth of the AT&T toll system, 1890–1904. Source: Abler, 1977, p.333.

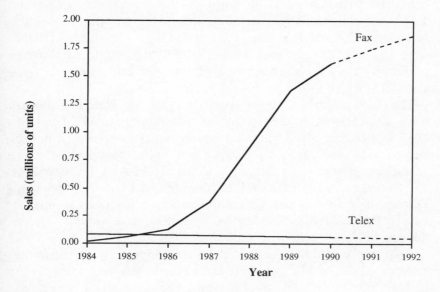

Figure 3.6 Sales of facsimile machines and telexes in Europe, 1984–92. Source: *The Economist*, 1990, p.27.

made fax a cumbersome, slow and expensive device, used mainly for military, meteorology and press purposes. Several developments as of the early 1960s turned the fax machine, 20 years later in the mid-1980s, into standard business equipment world-wide: various countries permitted the penetration of fax into the telephone system; international standards for scanning and transmission were agreed; digital technology was applied to fax transmission, and the machines became smaller, cheaper, faster, and more sophisticated.

Given these qualities, it seems that no other telecommunications technology diffused faster than the digital fax. The machine did not require any new networking, it further did not require any new skills, and it solved an obvious need to overcome slow postal services, as well as the slow, noisy, and text-only telex technology. It made the transmission of written information as instantaneous as the spoken one. Within eight years only, the sales of fax machines in Europe have gone through a complete diffusion cycle, in form of the logistic curve (Figure 3.6). A year after its introduction, sales of fax machines surpassed those of the older telex technology. In the US, there were 250,000 fax machines in use in 1985, growing to some four million by 1990, or a growth rate of 1600 per cent in merely five years! (Martin, 1991).

The hierarchical diffusion pattern typifying telecommunications innovations, applies to the fax technology as well. Thus, in 1987 44 per cent of all fax machines in Israel were located in the city of Tel-Aviv, and only 6.7 per cent were located in smaller Jerusalem. Three years later, the share of Tel-Aviv decreased to 35.0 per cent, while that of Jerusalem increased to 7.1 per cent. During these three years the total number of fax units in operation in Israel, increased from 990 to 12,827, or a growth rate of almost 1200 per cent!

Like the telephone, the fax was first adopted by the business community. It is intriguing to see whether this technology will diffuse into households as well, once the problem of unwanted incoming messages will be solved, and machine prices will continue to decline.

Home computers

The relatively low household penetration rate of home computers in the US and the high one in the UK were mentioned before. The US sales figures present a complete and fast diffusion cycle occurring between 1976–83 (Figure 3.7). However, this cycle points to a probable high penetration of micro-computers to businesses, since we have noted before a low one for households. Thus, the ups and downs in sales since 1983 may reflect the introduction of new models, as well as the general business climate.

The diffusion of computer telecommunications for personal business use has been remarkable. In 1971 Citicorp was the first bank that offered ATM (Automatic Teller Machines) services, and by 1987 more than 80 per cent of the customers used ATMs for over half of their transactions (Price and Blair, 1989, p.130). By the same token, the diffusion of the BITNET inter-

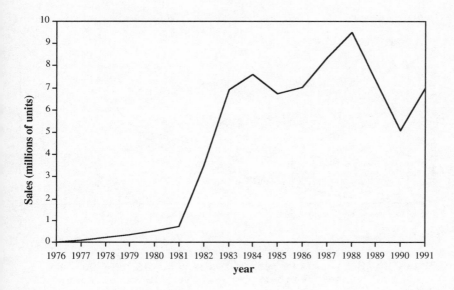

Figure 3.7 Annual sales of personal computers in the US, 1976–91. Data sources: Toong and Gupta, 1982; U.S. Bureau of the Census, 1991; *Ma'ariv*, 1992a.

university computer network diffused within three-and-a-half years of its introduction in the US to a point of near saturation (Kellerman, 1986a; 1986b; see also Lewis, 1989).

Cable television

Cable television was commercially introduced in 1951 (Table 2.1). It was invented in 1948, when a television technician in rural Pennsylvania, John Walson, built an antenna on a mountain which blocked reception of programmes in his area, and connected local residents to the antenna with cables. For over a decade cable television was used in the US, and even more so in Canada, for this purpose only. Thus, cable television presents a unique form of diffusion in its early diffusion phase, from rural to urban areas. In the mid-1960s, especially since the introduction of satellite-transmitted broadcasting, special cable

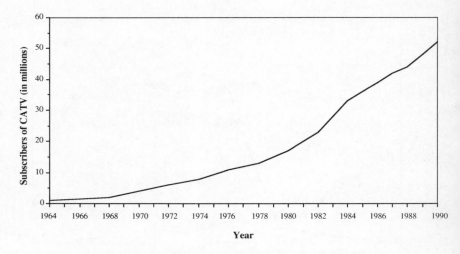

Figure 3.8 Subscribers for cable television in the US, 1964–90. Data sources: *Television Factbook*, 1982; US Bureau of the Census, 1991.

programming started. The diffusion of cable television in Europe has been slow relative to the US, and it was introduced in the UK only in 1980. However, even in the US the technology is far from a universal household penetration.

Cable television was available in 1990 to 90 per cent of US households (McCarroll, 1991), but only 55.7 per cent of US households, or less than two-thirds of the potential subscribers chose to buy the service. This is further demonstrated by the diffusion curve, which has developed slowly but persistently since 1964, and is still far from saturation (Figure 3.8). This slow pace has obviously to do with the monthly payment required for the service, the wide choice on regular channels, and the rapid diffusion of VCRs. Early in 1992, the FCC permitted the use of radio-waves for consumer response to cable programming, so opening a new channel for two-way cable television, especially for retailing (*Ma'ariv*, 1992b).

A unique form of television services is teletext, permitting the air transmission of texts to subscribers using decoders. This technology has been met with much success in the UK, as well as in Sweden, Finland and Israel. However, it has not penetrated the American market (Carey and Moss, 1985).

Telecommunications versus transportation: substitutability and complementarity

The use of telecommunications means may potentially constitute either substitution or complementarity to transportation means. As a substitute, it may eliminate the user's need to physically move from one point to another, whereas in the second option the use of telecommunications is assumed to bring about further travel (Salomon, 1988). Both options have been shown to exist in various spheres of human activity, though substitution has been limited by human desire for mobility (Salomon, 1985), as well as by human tendencies for direct sensing and socializing. Economically, transportation means are more distance sensitive than telecommunications, while the latter is more time sensitive than transportation (Brooker-Gross, 1980).

Telecommunications may potentially replace or complement travel for four major functions, and at different geographical scales: commuting and shopping at the urban level; business meetings at the urban and even more so at the regional level; and conferencing at the regional, national, and even international levels.

Telecommuting

It has been estimated that only 16 per cent of urban travel is for the purpose of transporting goods (Brooker-Gross, 1980). Some 42 per cent of urban travel, or one half of the motorized movement of people about the city is for commuting (Nilles *et al.*, 1976). A full 'domestication' of information work predicted by futurists (e.g. Toffler, 1981) could potentially alleviate traffic congestion, typifying cities world-wide. In reality, *flexispace* or *flexiplace*, namely the performance of some work at home has been more common, though even this option is still quite limited in its extent (Kellerman, 1984). Thus, in 1986–87, some

15–16 per cent of the US workforce performed *some* work at home, while only 1.6–2 per cent worked at home as their primary employment location (Kraut, 1989). In 1989, some 11.9 per cent of the US workforce worked on a *flexitime* schedule, namely that they enjoyed flexible working hours (US Bureau of the Census, 1991). It was also found that trip frequency for information workers in large American cities declined between 1977–83 (Kumar, 1990).

Workers at home may be classified into five groups (Olson, 1989): First, 'the privileged worker', or those who like to work at home and have the proper conditions to do it. Second, those who are forced by their employers to work at home, especially part-time female workers. Third, those whose employers have institutionalized working at home. Fourth, self-employed home-based workers, and finally, 'after hours' home workers. It is possible to remotely control work at home through computers, and it might be cheaper for an employer to invest in home-computers, modems, etc., rather than in office buildings. However, as repeated experiments have shown, most workers like the formal and informal socializing at the work place, in addition to the needed separation between domestic and work responsibilities, preferred mainly by women (Salomon and Salomon, 1984).

Teleshopping

Telecommunications-based shopping implies the use of videotex or two-way cable television for visualized shopping, or the use of the telephone for shopping of items advertised in various media. All these options are currently in operation, though two-way cable television is mostly experimental. In addition, super-market chains in many countries offer ordering services via the telephone, fax, computer and answering machines. In the early 1980s, direct marketing in all channels in the US amounted to some 12 per cent of total retail sales (Kellerman, 1984). In The Netherlands, mail orders amounted to 5 per cent of the consumer market. About 40 per cent of these mail orders were for clothing and some additional 15 per cent were for books (de

Smidt, 1991). Higher levels of education and income typify the clientele of teleshopping services (de Smidt, 1991; Koppelman *et al.*, 1991).

Store shopping has several advantages over teleshopping, and some of these advantages resemble the advantages of working places *versus* telecommuting. Store shopping permits a multisensory exposure to merchandise, which may be important psychologically and may as well permit price comparison. Store shopping is, furthermore, a form of recreation and pleasure. Teleshopping, on the other hand, is time-saving and time-flexible (Koppelman *et al.*, 1991). Thus, it seems that in the foreseeable future, teleshopping will still be complementary to store shopping rather than substituting it (de Smidt, 1991).

Business calls

It has been estimated that the use of the telephone for business calls replaced 20–60 per cent of local business travel (Nilles *et al.*, 1976), 34 per cent according to another study (Goddard and Pye, 1977), and 60 per cent according to a third one (Christie and Elton, 1979). On the other hand, it has been repeatedly argued that the telephone serves for the organization of meetings rather than substituting for them, and that telephone calls and meetings reinforce one another (Harkness, 1973; Brooker-Gross, 1980; Price and Blair, 1989).

At the interregional level, a study of 6000 business travellers between the north and south-east of England revealed the following trends (Goddard, 1983, p.121):

> The greatest substitution potential was to be found in the case of meetings held by representatives from multi-site companies compared with independent companies, within the manufacturing sector compared with the service sector, and taking place in the Northern Region compared with the South East. Communications by branches frequently involve research staff on matters internal to the organization, and for these communications the company has the facility of installing compatible equipment throughout the organization. In contrast, communications generated by independent firms frequently involve senior staff in negotiations with customers

and suppliers who may not have compatible facilities.

Teleconferencing

We referred in Chapter 1 to videoconferencing technology, noting the costs and limitations involved in its application. Commenting here on the possible substitution of travel by telecommunications, it is interesting to note the findings of a study comparing air travel, telephone conferencing and various video conferencing technologies (Salomon *et al.*, 1991). The study compared costs only, and could not quantitatively evaluate the benefits of face-to-face meeting *versus* the use of telecommunications. The study measured the costs of conferencing in 1988 at varying distances from Chicago. As the authors noted, telecommunications technologies improved in the meantime and the costs of their use declined, while those of air travel increased.

Taken into account were variables such as the varying numbers of participants, length of conferences, distances among participants, and costs for telecommunications or for travel and lodging. Time barriers in form of common working hours could not be taken into account. Generally, the use of the telephone turned out to be the cheapest medium, travel was intermediate and video was shown to be the most expensive medium. However, the costs for the latter declined with distance more than the two other modes of communications. Also, 'the larger the number of participants the more attractive is telecommunications, and the longer the meeting, the more attractive is travel. In addition, longer distances favor telecommunications while shorter distances favor travel' (Salomon *et al.*, 1991, pp.314–15).

Conclusion: the telecommunications cycle

We may summarize the various aspects discussed in Part 1 through the introduction of 'the telecommunications cycle' (Figure 3.9). The cycle begins with two major components,

Spatial dynamics of telecommunications

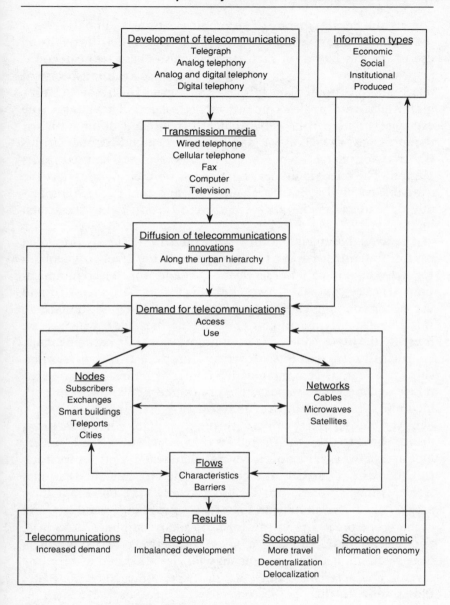

Figure 3.9 The telecommunications cycle.

namely the development of telecommunications technologies and the types of information which may be transmitted through the use of devices based on such technologies. These two categories respectively represent the hardware and software aspects of telecommunications. The information types consist of the four major classes for the contents of messages. They imply the existence of numerous geographical origins and destinations of the messages, as well as varying levels of demand for the various transmission media, depending on the type of transmitted information. The telecommunications media, on their part, reflect the available technologies at a given time, and they are geographically dependent on the phase of their diffusion along the urban hierarchy.

The available media, jointly with the specific needs presented by the four information types, determine the levels of demand for telecommunications services. Demand for telecommunications services consists of two components, namely access to and use of telecommunications systems. Access relates to the passive side of telecommunications, or incoming messages, whereas use presents the active side of telecommunications, or outgoing calls. As such, the two aspects of demand are in a two-way relationship with the telecommunications infrastructure consisting of nodes and networks, since these nodes and networks serve both incoming and outgoing traffic.

Every message transmitted through any telecommunications device has to be channelled through networks and nodes. Optimal networks may offer several channelling options for each call, and this is usually the case for intercity and international transmissions. Nodes, on the other hand, may present a hierarchy of telecommunications foci, switching centres, and exchanges (sometimes in the form of smart buildings, teleports, and telecottages). The flow of messages between two subscribers in telecommunications systems may be classified and assessed by several characteristics, such as distance, time, direction, etc. Flows may further be restricted by several barriers, such as regulation, time and language.

Geographical as well as other results attributed to information flows through telecommunications are varied, and they will be highlighted in various chapters of Part 2. For telecommunications *per se*, access to and use of the telecommunications system may bring about additional uses as well as additional exposure to potential contacts, or increased levels of demand. These would start new rounds of the telecommunications cycle. At the regional level, we have noted the accumulation of controlling power in larger cities, as a result of the introduction of telecommunications means, while peripheral regions may attract industrial plants controlled from the cores.

From a sociospatial perspective, access to and use of the telecommunications system may imply more travel triggered by exchanges of messages. Telecommunications may permit a decentralization of social and family ties, and a resulting lower attachment to specific locations, or a delocalization process. At the socioeconomic level, the use of telecommunications is a routine activity of the information economy, but the accumulation of such activities may sustain the development and specialization of the information economy.

The process described so far is cyclical, since each of these results may have some impact on previous phases of the process. Growth in the information economy could bring about a continued specialization in the four information types. The centripetal and centrifugal impacts of telecommunications may develop higher levels of demand for telecommunications services by all types of users, whether households or businesses, at both cores and peripheries. Such new demands may, on their part, exert pressures on existing technologies, so that new improvements and innovations may be sought. The telecommunications cycle was shown to start at the macro levels of technology development and information types, and gradually moving to the micro level of individual users. The aggregate telecommunications behaviour of users leads to macro-level flows and results, and these have repercussions on the macro levels of information types and technological improvements and innovations.

PART 2
TELECOMMUNICATIONS IN GEOGRAPHICAL CONTEXTS

Chapter 4
Telecommunications and Cities

The important role telecommunications may play in the spatial functioning of cities as well as in their urban economies has been recognized relatively early, for example by Christaller (1933) and by various studies in the 1950s (see Pred, 1973). Many cities, notably European ones, have recently attached telecommunications symbols to their identity. Manchester would like to become Europe's first 'on-line city'; Barcelona calls itself 'Telematics City'; Cologne has used the label 'Communications City'; and Amsterdam declared itself 'Informatics City' (Hepworth, 1990, p.196).

The interaction between telecommunications and cities received quite wide attention concerning the impact of cities on the development and patterns of telecommunications, on the one hand, and the impact of telecommunications on various aspects of urban systems, on the other. These two sides of the coin will be the focus of the discussions in this chapter. The possible impact of cities on telecommunications will be presented through the emergence of a global urban system, specializing in capital markets, information distribution, and in the controlling and management of industrial production and service activities. The impact of telecommunications on cities will be explored through the same aspects, as well as by an analysis of decentralization of back offices and data processing in metropolitan areas. Preceding these two major parts of the chapter will be an elaboration on the question whether telecommunications may lead to urban concentration or dispersal.

The approach which will be put forward throughout this chapter is that telecommunications technology may be considered as spatially neutral, namely that it can potentially bring about both enhanced concentration and decentralization. Which patterns will eventually prevail and to which degree depends on a myriad of other factors, such as economic specialization, city size and location, and the phase of the diffusion process of emerging technologies. The discussion will focus on actual patterns, and will avoid, as much as possible, futuristic treatments of telecommunicated cities.

It is of interest to note a narrowing down tendency concerning the issues regarded as relevant for a discussion of telecommunications in an urban context. In the 1970s, a conception was maintained that telecommunications may affect all aspects of urban life, but more recently attention has focused on business or producer services as being both the most affected sector by telecommunications, as well as the sector most affecting telecommunications systems (see Kellerman, 1984). We noted before the rather limited effect of telecommunications as far as telecommuting, teleshopping and teleconferencing are concerned. As will be shown, the interrelationships between the business sector and telecommunications may lead to the concentration of business services in larger cities side by side with a decentralization of some of these services within them. Both processes initially emerge at the highest echelons of the urban hierarchy, and move later to lower levels of the urban hierarchy.

Spatial concentration or dispersion?

The question whether spatial concentration or dispersion takes place in an urban context may refer to two possible processes, interurban and intraurban. At the interurban level, the question relates to two possible yet opposite options. A decentralization of economic activities may occur, mainly in manufacturing and routine service activities, moving from major metropolitan areas to smaller cities and towns, or a possible increased concentration of such activities will develop in large metropolitan areas. From an intraurban perspective, dispersion refers to the potential decentralization of economic activities as well as residences within urban areas. One gets the impression that these two processes are sometimes confused with each other, or simply not well differentiated. As current circumstances have proven, the interurban and intraurban dispersion processes which actually take place are related to each other, or better to say, intraurban decentralization resulting from developments in telecommunications may be induced by processes of interurban dispersion. The dispersion of industrial production from large cities to peripheral areas, aided by telecommunications means, amounted to an increased controlling power of the larger metropolitan centres. This increased power, which may also be coupled with enhanced global activities in large cities, may bring about higher demands for real estate in CBDs. These demand pressures may also be relieved by telecommunications technologies, since telecommunications enables back offices to suburbanize and still be controlled from CBDs. These processes have been rather gradual and non-revolutionary, but as we will see initial observations and forecasts proposed much more radical trends.

The antipolitan approach

Early reflections on city form and functioning under modified telecommunications technologies suggested a total dissolution of cities at large, or at least of CBDs. Gottmann (1977) termed such a dispersed settlement as 'antipolis'. Various versions of antipolis or 'the wired non-city' were put forward in the 1960s and 1970s (Webber, 1968; Goldmark, 1972; Short et al., 1976; Lehman-Wilzig, 1981; Abler, 1970), and their increased popularity made them appear as 'conventional wisdom' (Moss, 1987b, p.535).

Generally, the antipolis is based on the assumption that 'the glue that once held the spatial settlement together is now dissolving, and the settlement is dispersing over ever widening terrains' (Webber, 1968). As far as the central city is concerned, this was supposed to imply that it 'is continuing to lose its locational advantage and uniqueness within the metropolitan spatial structure. Technological changes continue to lower the necessity for concentration. Emerging telecommunications technologies promise an even greater freedom of locational choice' (Kutay, 1986).

The introduction of improved telecommunications technologies should have meant a total change for metropolitan areas at large: 'For the first time in history, it might be possible to locate on a mountain top and to maintain intimate, real-time, and realistic contact with business or other associates. All persons tapped into the global communications net would have ties approximating those used today in a given metropolitan region' (Webber, 1968).

As far as city systems are concerned the antipolis approach predicted that:

> Advances in information transmission may soon permit us to disperse information-gathering and decision-making activities away from metropolitan centers, and electronic communication media will make all kinds of information equally abundant everywhere in the nation, if not in the world. When that occurs, the downtown areas of our metropolitan centers are sure to lose some of their locational advantages for management and governmental activities (Abler, 1970).

Balanced views on the wired city

The antipolitan forecasts of total intra and interurban dispersal have not been realized, and an opposite trend could be observed *vis-à-vis* the concentration of controlling functions and global activities in large metropolitan areas. However, there is also a social side to the avoidance of radical dispersion in wired urban systems. Gottmann (1977, p.311) noted that the antipolitan view was based on several assumptions:

> . . . first, that access to the material transmitted by these means of information will fully satisfy most people for their work and leisure; second, that isolated living with good communications will satisfy most people; third, that the quality of personnel and the availability of adequate labor resources could be maintained by remote control; fourth, that the vast expenditure of energy and materials necessary to operate and maintain the networks of supply and the movements of a dispersed population would not be too costly; and last but not least, that most individuals have no reason for frequent recurrent presence in urban centers other than efficiency of work.
>
> None of these assumptions appears realistic.

Given the nature of urbanites and cities, many contemporary writers have concluded that the introduction of modified telecommunications means may bring about both concentration and decentralization affecting both population and economic activities at various scales (Wise, 1971; Nicol, 1985; Downs, 1985; Mandeville, 1983; Kellerman, 1984; Pool, 1990; Gottmann, 1977). Telecommunications has thus been seen as spatially neutral. It is supposed to present a permissive rather than a determinative impact, and whether concentration or decentralization takes place, for any given sector, depends on numerous other factors, such as transportation networks, labour mobility and more. As a matter of fact, even in the US, which has enjoyed a long and continuous expansion of its metropolitan areas, and has a long tradition of spatial expansion at large, no revolutionary spatial trends could be observed as a result of telecommunications developments.

The global urban system

A basic premise of urban geography is that cities are organized in national systems, consisting of functional hierarchies and various forms of rank–size orders. The development of telecommunications means has permitted the emergence of a global system of cities. Cities in this system function as integral components of their domestic economies, side by side with functions which they perform within a global economy of capital markets, producer services, and control of internationally dispersed industrial plants owned by transnational corporations (TNCs). The higher a city's rank in the hierarchy of this global urban system is, the larger the global component in its economy is and the smaller the domestic one becomes.

The following paragraphs will present basic terms for this global system, will trace its evolution, and will highlight the 'three-legged stool', topping the system, namely New York, London, and Tokyo. Obviously, the following discussions will accentuate the role of telecommunications in the functioning of the system.

General concepts

The most basic concept for the understanding of the global urban system tied together by telecommunications means, is 'the transactional city' (Corey, 1982; Gottmann, 1983). The transactional city specializes in the generation, processing, management, and transmission of information, knowledge and decisions, rather than in the production of tangible goods. Such transactions may be of a domestic or international nature, and both require a heavy use of telecommunications. Transactional cities may, therefore, be assessed in terms of their 'telecommunications distance', namely their global distance from national and international centres, as well as by their 'global location potential', or their capability to access other centres through telecommunications networks. Such capabilities are not only determined by the telecommunications distance, but also by the level of technical knowledge (Kutay, 1988). The emergence of transactional cities is an outgrowth of past industrial and international advantages, in addition to current high concentrations of interpersonal contacts in large cities. Thus, the global location potential of certain, but not too many, cities may be profoundly high.

The global urban hierarchy

Telecommunications distances, global location potentials, and the agglomeration of transactional activities have resulted in a four-tier global urban hierarchy, consisting of the following levels, from lowest to highest:

1. domestic cities
2. world cities
3. regional hubs
4. global hubs.

Before moving to a brief discussion of each of these levels, several comments on the hierarchy itself are in place. The global urban hierarchy should not be confused with a hierarchy of telecommunications nodes, which may consist of smart buildings, teleports and telecottages (see Chapter 2). The latter related to the spatial system of nodes in a telecommunications infrastructure, whereas the global hierarchy discussed here is an urban hierarchy based on transactions and telecommunications. The global urban hierarchy has some equivalence to the hierarchy proposed by Langdale (1987) for electronic information services (EISs), namely services which are collected, processed and transmitted electronically.

Domestic cities: These are cities, the urban economies of which are based on one or several of the following sectors: (a) manufacturing of goods of any kind; (b) tourism of all kinds, with a rather restricted share of business tourism; (c) transactions of information related to regional or national government, or information based on regional or national service economies. In other words, these are cities which make use of telecommunications, but in a more limited sense, as far as the international and business-controlling components are concerned. This observation means that domestic cities definitely constitute a component of a global economy of exchanges of commodities, people, capital, and information, but their role is more oriented towards the domestic, and to the routine aspects of the global economy. Also, these cities may tend to export more routine information, while being dependent on higher levels of the global urban economy, for the importing of non-routine information. This dependency may relate to numerous areas, ranging from business decision-making in large corporations, to capital market information, and television news and programming. Domestic cities may be of varying sizes, sometimes they may even be large metropolitan areas, and they are further organized within conventional hierarchies. Examples may include cities such as Baltimore in the US, or Leeds in the UK.

World cities: The term 'world city' was proposed by Friedman and Wolff (1982), and by Friedman (1986). It was sometimes referred to also as 'global city', or 'international city' (see e.g., Cohen, 1981; Knight and Gappert, 1984). Originally, it referred to all cities with a major international component, namely including regional and global hubs. Here we refer to world cities in a more restricted sense, so that it applies to cities with a major international component in their economies, simultaneously serving a region within a large industrialized nation, or serving a whole nation in the case of middle-sized or small industrialized or newly industrialized countries. As such, the major functions of a world city are as follows: (a) it hosts headquarters of transnational corporations of considerable size; (b) it hosts the domestically main branches of foreign transnational corporations; (c) it serves as a gateway for the served region or nation to global capital markets, as well as to international financial and producer services; (d) it offers sophisticated telecommunications services for the hosted functions and it serves as the major telecommunications junction for the whole nation/region. These telecommunications services, though, do not necessarily have to be offered in the form of smart buildings and teleports.

World cities may have their own hierarchies (Cohen, 1981; Moss, 1987b), and primary and secondary levels were proposed by Friedman (1986). For example, Paris may be considered a primary world city, whereas Copenhagen may be viewed as a secondary one, given the differences in the size and diversification of the domestically supporting territory, as well as the location of the city. Usually, world cities have been identified through the location of the headquarters of at least one *Fortune 500* corporation (see e.g., Cohen 1981; Knight and Gappert, 1984). World cities were further typified by their manpower, which consists of the highest professional levels of information and decision-making occupations, side by side with workers engaged in tourism and recreation, serving both domestic and foreign business tourists (Sassen-Koob, 1985). Furthermore, the transnational accent in the urban economy implies that the local culture and atmosphere accentuates openness to foreigners, and a blend between international and domestic cultural dimensions (Perlmutter, 1979).

Regional hubs: Regional hubs constitute world cities which serve several countries, rather than just one country or one region within a country as ordinary world cities do. Only a handful of cities function as regional hubs. Hong Kong and Singapore are prime examples, Bahrain may be a third one. In a certain way,

Brussels may also be considered as such, given its status as headquarters for the EC.

Regional hubs have the following characteristics: (a) they provide telecommunications services for neighbouring countries as well as for themselves; (b) they host producer services normally located in world cities, mainly banks, law offices, computer and advertising services, many of which are branches of foreign firms, notably American ones. All these services may be geared towards a clientele dispersed in several countries; (c) regional hubs are preferred locations for headquarters of transnational corporations (Langdale, 1987; Moss, 1987a); (d) regional hubs normally typify city-states which may, on the one hand, have an otherwise limited economic base, and which, on the other hand, are located close to other countries which cannot afford to develop world city facilities, for political and/or economic reasons. Hong Kong and Singapore have become attractive for global financial flows, because of their flexible policies in this regard (Castells, 1989, pp.339–40).

A country that could have potentially served as a nesting nation for at least one regional hub is Switzerland, which enjoys several inviting factors for the emergence of regional hubs. It is located in central Europe, close to both large and small European countries, it has enjoyed a neutral political status for centuries, and above all, it has developed a long tradition of international banking. There are several reasons why no Swiss city has developed into a regional hub despite these merits. First, the surrounding European nations, large and small, have their own urban, finance and telecommunications infrastructures. Second, Switzerland is a land-locked country, and one of the power bases of regional hubs is their maritime accessibility. Third, Swiss banking is conservative, serving to a large degree as a haven for 'silent' capital. Regional hubs, on the other hand, are based on aggressive entrepreneurial capital. Fourth, Switzerland, did not crucially need regional flows for its own prosperity and development, given its long traditions in industrial production (watches, foods) and tourism.

Global hubs: Three cities may be identified as global hubs, or the top level of the global urban hierarchy: New York, London and Tokyo. These cities may be typified by the following aspects: (a) they serve as world cities for a whole continent or sub-continent: New York for North America; London for Europe; and Tokyo for the Pacific Rim; (b) these cities are, furthermore, tightly interconnected among themselves, so that they serve as the top of the pyramid of the global urban hierarchy. This is reflected in their telecommunications facilities and traffic, as well as in their crucial role as decision-making and transactional nodes for global exchanges of capital, information and commodities economy; (c) the cities are largely tied to global economies, sometimes even more than to their own national or regional economies (Moss, 1987b); (d) their senior status in the global economy is not necessarily reflected in their population sizes, land area or employment, since these parameters may fit better industrial and urban hierarchies (Moss, 1987b); (e) they may develop global specializations in the handling of capital, so that they may 'function as one transterritorial marketplace' (Sassen, 1991, p.327).

This 'three-legged stool' of the global economy and global urban hierarchy (Hamilton, 1986) deserves more detailed attention, given its crucial global role, as well as the crucial role of telecommunications in its emergence and functioning. We will take a comparative look at these cities with the aid of Tables 4.1 and 4.2.

Table 4.1 The shares of New York, London, and Tokyo in global capital markets (in %)

Global hub	Share in intl. banking 1986	Share in world's stock markets 1985	Share of world's 100 largest banks 1988
New York	15	42.3	8.76
London	23	7.9	5.68
Tokyo	10	21.1	45.64
Total	48	72.3	60.08

Sources: Daniels (1991a); Hepworth (1991b); Sassen (1991).

London: London is the oldest international centre among the three global hubs; it received its status as a global hub, specializing in international banking, shipping, insurance, and trade back

Table 4.2 The shares of leading capital markets in the movements of people, commodities, and information (in %)

Nation	Telecom. Share in world's intl. telecom. 1990	Tourists Share in world's air passen. 1987	Tourists Airport rank 1987*	Trade Share in world's exports 1987
USA	25.0	2.8	3	10.2
UK	8.4	5.6	1	5.2
Japan	2.9	2.5	4	9.3
Total	36.3	10.9		24.7

* Airport ranking for New York, London and Tokyo, respectively.
Sources: Moss (1989); Staple (1991); US Bureau of the Census (1991); Daniels (1991b).

in the colonial era of the eighteenth and nineteenth centuries (Daniels, 1991b; King, 1990). Still the emergence of London as a leading global hub requires further explanation, because of the existence of some other potentially leading cities. Whereas New York and Tokyo have been unchallenged as the largest and predominant business concentrations in North America and the Pacific Rim, respectively, London could be challenged by cities such as Paris or Frankfurt on the continent. It was, however, London which became a preferred location for American transnational companies, and for a variety of reasons. It enjoys a closer location to the US, so that savings could be achieved in international telecommunications costs. This geographical dimension has been a major factor in the British move to privatize their telecommunications system in the mid-1980s and opening it to competition (Langdale, 1989a), and a factor in Irish attempts to make Dublin compete with London (Grimes, 1992).

London enjoys several other advantages as the European global hub. The English language is helpful for the American and international business communities. Also, the London financial market can do daily business in the morning with the still active Japanese markets, and conclude the day with interaction with the New York markets (Daniels, 1991b). The long tradition of international business in insurance, shipping, banking, and trade, have no doubt provided London with initial and cumulative advantages for contemporary computerized, telecommunicated, and diversified services (on London's functions in these areas see e.g. Thrift, 1987).

London is the global hub for capital processing (Sassen, 1991, p.327). It had a share of 23 per cent, or almost a quarter, of world international banking in 1986, and all three global hubs reached 48 per cent, with New York being second, and Tokyo third. London serves as an international banking centre, mainly for American and Japanese banks (Moss, 1987b; Daniels, 1991b). London further enjoys the most international stock market with 23 per cent of the listed companies being foreign (compared to 5 per cent in New York, and 7 per cent in Tokyo) (Langdale, 1992). The diversified nature of London's capital markets, the concentration of international shipping and insurance, the nature of international banking, as well as its location, have all turned London into the busiest city in the world, in terms of the passengers using its airports.

New York: New York is the second oldest global hub. It received its status in the second half of the twentieth century. However, whereas in the UK it was largely for special traditions which developed in London as a city that contributed to its status as a global hub, in New York it was for a blend of both prominent local and national circumstances that turned it into the world capital of stock markets and telecommunications.

Politically, New York became a 'world capital' with the location of UN headquarters in the city following World War II. It was, however, the rise of capitalism and the technological developments in telecommunications, both centred in the US, that played a major role in its becoming a global hub. The city enjoys being the centre of 'the world's largest market economy, the innovativeness of its financial and other institutions, its role as the center for the world's most widely used currency (the dollar) and its willingness to replace redundant structures with new ones' (Daniels, 1991a, pp.17–18). The comparative power of domestic capital resources may be demonstrated by pension funds. In 1985 US funds amounted to $1.5 trillion, compared to

$225 million in the UK, and $210 million in Japan (Castells, 1989, p.341).

New York's role in the 'three-legged stool' is its being the most important centre for capital investments, decision-making, and innovation of capital investment ideas (Sassen, 1991, p.327). New York's share in world stock markets in 1985 was 42.3 per cent, and the share of all three hubs was 72.3 per cent. This supremacy was coupled with that in telecommunications, notably in international telecommunications. Thus, the number of calls originating daily from Wall Street increased from 900,000 in 1967 to three million in 1987 (Warf, 1989). The share of the US in the world's international telecommunications in 1990 was 25 per cent, excluding leased lines. It was estimated that New York's share in US international telecommunications was 24 per cent before the divestiture of AT&T (Goddard, 1989), and that the combined share of New York and Los Angeles was 30 per cent (Moss, 1987a). For 1988 it was estimated that New York produced more than a third of America's international telephone traffic (O'Neill and Moss, 1991). Thus, it is possible to estimate that New York alone has a share of about 6–8 per cent of the public global traffic in international telecommunications (i.e. excluding leased lines). This share, though being high, is considerably lower than the city's share in stock markets. The difference may stem from the more restricted nature of stock markets selling stocks only, and at a small number of markets, whereas transmitted information is varied, and originates in many points. New York's supremacy in global telecommunications finds further expression in the introduction of new technologies first there. The first teleport was built there, and in the mid-1980s, over one-third of fibre-optic cables in the Bell system were installed in the New York area (Moss, 1986a).

New York is also the centre of the largest exporting nation, though Japan is not lagging behind much. These exports, and the striking seniority of the city in capital markets and telecommunications have attracted to the city domestic and foreign banking and securities (Moss, 1987b; Moss and Brion, 1991; Warf, 1991). 'Unquestionably the most potent force driving the New York economy is its banks; banking is to New York as the automobile industry is to Detroit or steel was to Pittsburgh'

(Warf, 1991). Forty-three per cent of foreign bank offices in the US were located in New York in 1986, with assets amounting to 59 per cent of their total assets. Leading were Japanese and British banks, a trend repeating itself in the other two nodes of the 'three-legged stool' of the global financial community.

The emergence of advanced long-distance telecommunications services and the deregulation of international banking have brought about a geographical change in US domestic banking as well, in the form of deregulation of interstate banking. This has turned New York into a stronger national banking centre, as well. Between 1982–90 the city gained $290 billion in assets as a result of interstate banking (compared to the second largest gaining city, Charlotte, North Carolina, with a gain of $85.3 billion).

Tokyo: The youngest joint in the 'tripartite axis' of the global finance economy (Warf, 1991, p.245), is Tokyo, which became part of the system in 1987, when it organized its offshore investment market (Castells, 1989, p.339). Tokyo represents the emergence of Japan as the largest source of international capital (Daniels, 1991a; Sassen, 1991, p.327), and the emergence of a geographically global economy, typified by flows of capital, information and high-tech products. Japanese investors purchased more than 40 per cent of US Treasury Department securities, and the Big Four Japanese investment firms account for 10 per cent of all stock transactions on the New York Exchange (Warf, 1991).

Tokyo hosted the headquarters of 11 of the world's 50 largest banks, and its stock market was the fastest growing one (Moss, 1987a). It represents Japan as the leader in the economic growth in the Pacific Rim at large, turning it into a third world economic core. It obviously represents tremendous growth within Japan itself, as a nation typified by high productivity and saving rates, coupled with aggressive technological innovativeness.

The development of the system

The evolution of the global urban system, as well as urban transitions as a result of its emergence are presented in Figure 4.1.

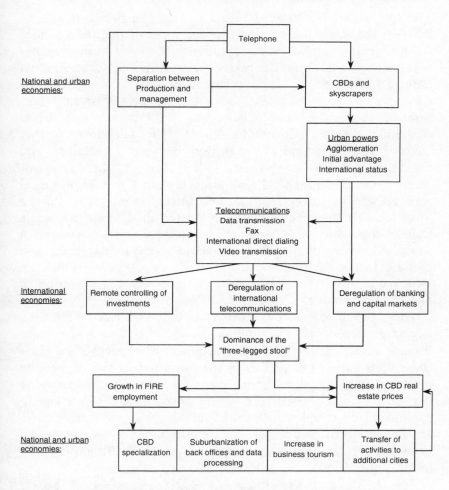

Figure 4.1 Telecommunications and business services in an urban context.

The two-way process is described as consisting of three phases. First, developments within urban and national economies, as of the end of the nineteenth century, when the telephone was introduced. Second, technological developments in telecommunications as of the 1970s, leading to the growth of international economies as of the 1980s. Third, developments in international economies and their urban spatial implications.

The following discussion will elaborate on the first two phases, whereas the third one will serve as the topic of the next section.

The introduction of the telephone as a major business communications tool implied two developments of urban-spatial nature. The telephone, as an information mover, jointly with the elevator, as a physical mover, contributed crucially to the evolution of skyscrapers and CBDs, the headquarters for urban economic activity (Gottmann, 1977). The telephone further enabled the geographical separation between production and management (Falk and Abler, 1980; Moss, 1987a; Goddard, 1989). Such a geographical separation pertained to both the local level as well as to the regional and interregional levels. At the local level, management could be located in downtowns, while industrial production took place in industrial zones elsewhere in town. At the regional and interregional levels, management and manufacturing could be located in different towns or regions. In both cases this separation implied more power concentration in and development of CBDs.

These developments have provided cities with differentiating agglomeration forces over many decades, so that larger cities became even larger in their production capacity, in their controlling power, as well as in the evolution of the supporting producer services, all of which were enhanced by the telephone (see also Daniels, 1991b; Fox-Przeworski, 1990). Nevertheless, agglomeration coupled with initial advantages gave certain cities an international status, in an international economy based primarily on exchanges of commodities, and using primarily telegraph and telex machines for their functioning.

The 1970s and the 1980s witnessed technological breakthroughs and widespread adoptions of computer data transmissions, fax communications, international direct-dialling, and video transmission. These were first introduced in large urban centres, thus reinforcing them even further (see also Fox-Przeworski, 1990).

These technological developments have meant a gradual elimination of temporal and geographical barriers for international transactions. In the 1960s telex replaced telegrams, so that interaction took minutes rather than hours and days. In the 1970s, direct telephone dialling permitted further temporal cuts, and was followed in the early 1980s, by on-time video transmissions of financial information provided by Reuters and other news services (Hepworth, 1991b). These technological improvements permitted, therefore, the formation of global financial and service economies (Daniels, 1991b; Moss and Brion, 1991).

The technical capability to remotely control business and investments was aided by two administrative-economic developments, namely the deregulation of international banking and capital markets, coupled with the deregulation of telecommunications. The deregulation of the financial markets has brought about various changes. In the mid-1970s Visa and Mastercard became international, and in 1977 an institutionalized international interbank electronic funds transfer system, SWIFT (Society of Worldwide Interbank Financial Telecommunications), started operations (Buyer, 1983; Langdale, 1985; see also Warf, 1989). Once again, larger banks in larger cities benefited from these changes more than financial institutions located in small and medium-sized cities (Moss, 1987a; Warf, 1991). In the US, five cities became major centres for foreign banking, in addition to New York: Los Angeles, Chicago, Houston, San Francisco, and Miami (Figure 4.2) (Moss, 1987b). These cities provide the US with a geographical belt of international banking centres, or international financial gateways, located along its maritime and continental borders. New York's weight is much higher than that of the other centres, and the number of foreign bank offices there is almost four times larger than that in the second centre, Los Angeles. In addition, the improvement in telecommunications and the emergence of global capital markets, permitted foreign locations for American law, accounting and advertising firms (Moss, 1987b).

The deregulation of international capital markets created a demand for more sophisticated and faster transmissions of financial information and funds (Warf, 1989). The 1990s have witnessed the possibility of introducing continuous round-the-clock global trading in securities. Demands for such services are not yet high, and the major financial markets expect a too harsh competition among them. However, various scenarios have been designed, such as extending the operating hours of major

110 Telecommunications & geography

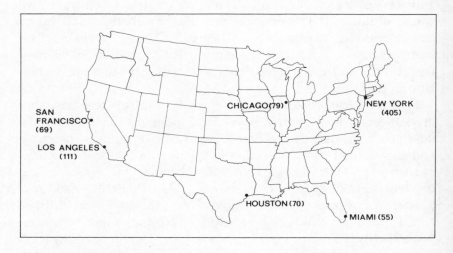

Figure 4.2 Major centers of foreign bank offices in the US, 1986.
Source: Moss, 1987b.

markets, using electronic trading, and letting a trading company move its 'book' from one market to another 'following the sun' (Langdale, 1992).

The deregulation of international and domestic long-distance communications effected larger cities first and in several ways. First, deregulation was first introduced in the largest cities, coupled with the introduction of technical improvements, such as fibre-optics. Second, the cancellation of cross-subsidies aided large cities, since the larger volumes of messages which they generated permitted lower basic rates (Daniels, 1991b). Third, large companies with high demands for telecommunications services could negotiate for lower rates (Daniels, 1991a). It is interesting to note in this regard that the first two countries that followed the US deregulation of long-distance telecommunications were the UK and Japan, namely the two other countries hosting global hubs, and so requiring competitive and sophisticated international telecommunications systems.

The location of producer services

The growth in producer services in large cities, notably in the international sector, may be expressed in two ways: rising prices for real estate in CBDs, and growth in the number of those employed in the FIRE sector. When these trends are coupled with enhanced telecommunications means they may lead to three spatial developments: suburbanization of back offices and data processing; CBD functional specialization; and transfer of service activities to other, usually smaller, cities.

These developments imply both concentration (in CBDs) and decentralization (suburbanization and transfer), both within metropolitan areas and in the urban system at large. Such possible simultaneous contradictory spatial effects of telecommunications were noted already by Gottmann (1977). It is further possible that no significant geographical change will be identified as a result of the developments portrayed in the previous section. This may notably be the case in small countries, and if the rate structures of telecommunications services are not flexible enough (Nijkamp and Salomon, 1989). Another possibility is that certain services will have different locational trends in cities of varying sizes.

Office suburbanization

Telecommunications does not constitute the only reason for the suburbanization of back offices and/or data processing facilities (Moss and Donau, 1986). Moreover, one could also reason that 'telecommunications do not directly cause decentralization, but they create the opportunity to make a decentralization decision' (Kutay, 1986). There are various direct causes for the suburbanization of offices, the most important of which might be higher inner-city rents (Goddard and Pye, 1977), and as we have noted, higher rents may be indirectly related to enhanced telecommunications. However, when considering the office relocation issue, a wider 'contextual' view of pros and cons has to be taken (Daniels, 1987). A high percentage of female suburban employees, especially part-time ones, may call for suburbanization of back offices, so that commuting time and costs would be cut. This factor has to be weighed, however, against existing transportation infrastructures and travel habits. Suburbanized offices require their own accompanying physical infrastructure (such as shopping malls), as well as a positive attitude by planning authorities. Office relocation options may be offset by geographically conservative investment policies by banks, as well as by the merger of companies willing to stay in older locations. Furthermore, services, notably producer services, tend to suburbanize following, or in conjunction with other, similar or different services, thus demonstrating high levels of urbanization and localization economies (Kellerman and Krakover, 1986; Howland, 1991).

The double trend for centralized decision-making and dispersed routine data processing could be traced in the US back in the 1960s (Moss, 1987b). However, in 1984 central administrative office employment was still quite centralized. About 42 per cent of such jobs in the US were located in the core counties of 24 of the largest metropolitan areas (Drennan, 1989). Those who moved used, however, intensely sophisticated telecommunications means (Kutay, 1986), reduced the average commuting distance and frequencies, and were able to elevate the quality of working life (Kumar, 1990).

Following a survey in Pittsburgh, Kutay (1986, p.248) was able to trace the sequence of office decentralization:

1. Diseconomies of downtown initiate the search for a new location.
2. Technological advances increase the effectiveness of telecommunication systems, thus making information transfer via networks and communications channels an attractive substitute for business contacts, reducing the agglomeration economies that tie office activities to the CBD, and creating the opportunity to decentralize.
3. Beginning in larger sophisticated firms (which are more influenced by telecommunications technology in their locational decisions), office activities no longer tied to the CBD are decentralized to non-CBD locations.
4. Organizations which depend on national and international

markets and/or are less conservative in terms of employing new technologies have a greater tendency to select a suburban location.
5. Firms which are organizationally decentralized and have delegated the authority to business units and divisions decentralize those units to suburban areas where land is cheap and there is more space for expansion.
6. In the case of partial decentralization of headquarters activities, departments which are more autonomous and self-sufficient and depend on telecommunications for contacts with other departments decentralize to suburban locations to overcome the diseconomies of the CBD.

Of special interest is Kutay's finding that companies with a national or international orientation would tend to suburbanize, whereas Noyelle and Peace's (1991) observations for New York seem to suggest the opposite. The type of suburbanizing head offices may depend, therefore, on the city's ranking in the global urban system. Cities ranking high depend more on the international component in their economies, so that companies headquartered there would require downtown locations for both face-to-face contacts and advanced telecommunications facilities. On the other hand, domestic cities depend more on local and regional economies, so that face-to-face contacts and prestigious locations are demanded more by companies geared towards the local and regional markets (as was found by Kutay's (1986) survey).

CBD specialization

CBD specialization is the other side of the coin of decentralization of back and/or front offices. Firms which remain in CBDs with higher real estate rental or purchase costs are companies which require core and prestigious locations, or companies the specialty of which fits the changing urban economy. The first case would occur mainly in domestic cities, whereas the second would fit world cities. CBDs would attract internationally oriented companies dealing with the movement of commodities, information, people, and capital. They would further attract firms dealing with information production and transmission, such as newspapers and broadcasting companies, as well as telecommunications and computer companies (Kellerman, 1984).

In Los Angeles the very emergence of a CBD was related to the development of the city as an international capital market, especially for dealings with the Pacific Rim (Moss, 1987b). In the metropolitan area of New York, the central city gained between 1977-84 87 per cent of the metropolitan employment increase in securities, 65 per cent in banking and 59 per cent in legal services. Gains for other producer services were higher in the suburban rings, reflecting growth in local and back-office employment (Schwartz, 1992).

Business tourism, notably the international one, increases with the growth of the global economy. It may create its own demands for telecommunications services, both before and during visits. This movement of people is considered the fastest growing component of the international service trade (Warf, 1991). Business tourists in world cities could, thus, bring about a reinforced specialization of CBDs in high-class business hotels.

Transfer to other cities

Rising real estate and labour costs coupled with advancing telecommunications networks may bring about the transfer of service activities from higher ranking global cities to lower ranking centres. Back in 1977 Goddard and Pye could not see such relocation happening in the UK from London to points beyond the South East. However, the US experience shows such transfers taking place especially from the highest ranking city, New York.

Such relocations may take one of two forms, namely a complete relocation of a firm or its geographical separation. Relocation had its roots as far back as the 1960s, mainly in nationally oriented companies in the food, airline and retail business. Recipient cities were regional centres, such as Atlanta, Boston, Minneapolis, and Dallas (Noyelle and Peace, 1991). A geographical separation may leave a firm's headquarters, in

New York, while data processing facilities are moved to lower ranking urban centres. Communications with headquarters is maintained via sophisticated, sometimes even specialized, telecommunications networks. Striking examples are the transfer of Citibank credit card divisions to Tampa, Florida, and Sioux Falls, South Dakota; American Express moving its processing facilities to Salt Lake City, Phoenix, and Fort Lauderdale; Hertz moving to Oklahoma City, and Avis transferring to Tulsa (Warf, 1991).

Other attempts to move information activities down along the urban hierarchy are the planned construction of industrial and service parks based on advanced telecommunications technologies and networks in various countries. Notable are the Australian multifunction polis (MFP), the Japanese and French technopolis, and the German media park (Langdale, 1989c; Morgan and Birley, 1990; Hepworth, 1991a).

Conclusion

Telecommunications maintains a strong two-way relationship with cities as spatial and economic entities. However, rather than bringing about a radical change in form of large-scale intraurban and interurban dispersion processes, telecommunications was integrated into the complex web of urban life in more restricted and discriminating ways.

The introduction of modern telecommunications means has accentuated the role of cities as service and control centres. Telecommunications means have thus contributed more heavily to larger cities, the scope of activity and influence of which has growingly consisted of a strong and developing international sector. Suburbanization and relocation of service functions may be facilitated through telecommunications services, by connecting smaller centres to larger ones. However, the very geographical move of service functions is dependent on various other factors, which on their part may reflect earlier impacts of telecommunications. A prime example is higher real estate costs in CBDs, resulting from the concentration of controlling functions there.

Chapter 5
Telecommunications and Regional Development

The introduction of well-developed telecommunications means may potentially result in growth of service and manufacturing activities in peripheral regions. Such growth implies, however, a separation which we noted earlier, namely that between controlling activities bound to major metropolitan centres, and production of goods and services in the peripheries. This vertical disintegration is similar to the intraurban separation between front and back offices (Warf, 1989). It further demonstrates the role of telecommunications as facilitating the emergence of new spatial divisions of labour, similarly to transportation, which may eventually yield new locational specializations. The following sections will, first, outline the role of telecommunications in the process of regional development, followed by elaborations on the major developing economic sectors. These two discussions will be complemented by an exposition of the trends and patterns in several major peripheral regions.

The role and context of telecommunications in regional development

Telecommunications has been recognized as an important dimension for contemporary regional development. It is, thus, in place to outline, first, the role and context of telecommunications in the development process at large, before moving to the specifics of the leading economic sectors in the development process, namely manufacturing and services.

The role of telecommunications

At a first glance, telecommunications seems to be a major 'factor' for economic development, or in other words, the very introduction of advanced telecommunications services into peripheral rural and urban settlements and regions would bring about economic development, since it would permit efficient communications between cores and peripheries. However, as Table 5.1 demonstrates, many scholars view the role of telecommunications in more modest terms. It is generally seen as a necessary vehicle for development rather than as a factor, implying that a variety of other conditions have to be met, and accompanied by supportive policies, in order for telecommunications to contribute to the development process.

Table 5.1 The role of telecommunications in regional development

Role	Sources
Enabling	Hepworth, 1990, p.68; Abler, 1991, p.44.
Making possible	Abler, 1977, p.331.
Necessary but insufficient	Gillespie, 1987, p.230; Salomon and Razin, 1988, p.123.
Offering new opportunities	Robinson, 1984, p.361.
Permissive	Goddard and Gillespie, 1986, p.392.
Catalytic	Parker et al., 1989, p.3.
Factor of production	Saunders et al., 1983, p.73.

Telecommunications constitutes, therefore, a basic condition for development, like transportation infrastructure, but its very existence in a region does not yet guarantee regional development. Also like in transportation, telecommunications facilities may, in principle, assist developments in a wide variety of economic activities (Parker et al., 1989, p.3). Telecommunications further permits faster and more efficient flows of information from cores to peripheries, both to businesses and households (Dillman, 1985; Robinson, 1984).

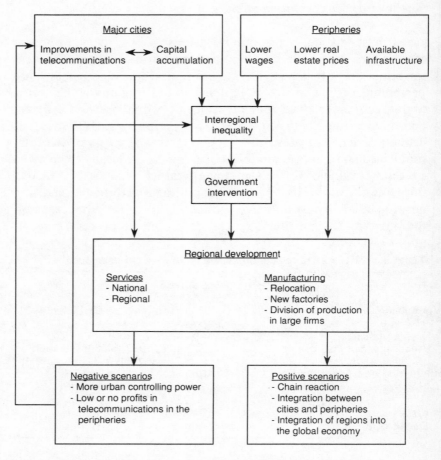

Figure 5.1 Telecommunications and regional development.

Telecommunications and interregional inequality

At least five aspects are involved in telecommunications-related regional development (Figure 5.1). Of these, two relate to the urban cores (improvement in telecommunications and capital accumulation) and three to the peripheries (lower wages, lower real estate prices, and available infrastructure). Telecommunications systems in major cities are assumed to be advanced to a level that permits control of remotely located activities. The development of a telecommunications infrastructure in major cities is related to capital accumulation. On the one hand, capital accumulation may call for additional production of goods or services in more efficient ways, namely in locations which offer lower production costs. On the other hand, we noted that contemporary capital accumulation is also dependent on communications means for efficient investing. In addition, telecommunications *per se* is a business that calls for technological advancement in order to permit capital accumulation in it (Martin, 1991).

The peripheries may become attractive for business if wages in the pertinent occupations are significantly lower than in major urban centres (Howland, 1992). By the same token, the cost of real estate in the periphery should be significantly lower than in major centres. Furthermore, proper infrastructures, notably transportation and trained manpower have to be available in the periphery (see Hansen *et al.*, 1990; Batty, 1988). Normally, regional development in the periphery becomes possible only with governmental intervention. This may take the form of assistance for enhancements in telecommunications services in the periphery, as well as the provision of extensive connections to major urban centres. Governments may further assist peripheral regions by training human resources for employment in telecommunications-related activities, and by making available transportation facilities. Sometimes, direct subsidies for businesses are required. Telecommunications-related regional development may occur in manufacturing and/or services, and we will detail both sectors in the following section.

Telecommunications-related regional development may involve additional developments of a positive or negative nature. A positive scenario calls for a chain reaction, so that additional factories and service functions be established in the periphery, both heavily and lightly telecommunications-dependent ones (Hansen *et al.*, 1990). Regional infrastructures may also be upgraded, notably transportation facilities (Nicol, 1983), and enhanced telecommunications systems could further imply a better integration between cores and peripheries if more contacts would emerge (Gillespie, 1987). At a later stage, careful development efforts, particularly of telecommunications services, may lead to an incorporation of peripheral regions into the emerging global economy (Gillespie and Williams, 1988). All these positive developments imply the existence of provincial cities in peripheral regions. They possess some earlier telecommunications advantage (Goddard and Gillespie, 1986), as well as an institutional infrastructure which permits them to take a leading role in the development process (Robinson, 1984).

A negative scenario suggests that the core cities will gain more than peripheral regions from the development of the latter. The controlling power of the core cities would increase, thus calling for the introduction of more powerful telecommunications services, so that interregional inequality would rise rather than diffuse (see Figure 2.1). Furthermore, low growth rates in demand for telecommunications services in the periphery may slow down additional developments by service providers (Hansen et al., 1990). Such circumstances may involve imports of cheaper services from the cores, especially to weaker regions, so that regional inequalities will increase (Gillespie, 1987; Beyers, 1989).

As is evident from the discussion so far, telecommunications is tightly integrated with numerous other dimensions and factors of development. A possible success of telecommunications-based regional development calls for a detailed and careful examination of specific local weaknesses and strengths (Gillespie, 1987).

Economic sectors

Telecommunications may be useful for both manufacturing and services, especially for producer services located in peripheral regions. In both activities, the transmission of information is the striking dimension of telecommunications, accentuated in producer services, where telecommunications permits the very production and processing of information in the periphery. The type of transmitted information, as well as the skills of workers involved might obviously be different than in manufacturing.

Manufacturing

Factories opened in peripheral regions in connection with improved telecommunications facilities and services may be of several types. First, they could constitute outlets of firms headquartered in metropolitan areas, using telecommunications to control factories in the periphery. Such controlling could range from the transmission of accounting information from the factory to company headquarters, to remote CAD/CAM (computer aided design; computer aided manufacturing) from the headquarters to outlying factories. Goddard (1983) and Hepworth (1990, p.94) claimed that remotely monitored outlying industrial plants would constitute most of the new factories established in peripheral regions as a result of telecommunications improvements. However, new factories may also be of other types, such as new independent firms or older independent firms relocating from leading industrial centres. Relocating firms may enjoy lower production costs in developing regions, while being able to conduct business with urban centres within and outside the country (Hansen *et al.*, 1990).

Early evidence for the first option was provided for Italy by Antonelli (1979), whereas the two additional options were described by Sheffer (1988) for high-tech firms in Israel, and for Bavaria, Germany by Genosko (1987). Genosko noted that major urban centres are typified by larger industrial plants compared to peripheral regions. Such large factories have a higher tendency to use telematics, given their better qualified workers, their openness and willingness to introduce new communications means, as well as better access to capital markets.

Services

One may find extreme differences in the assessment of the role of telecommunications in the development of services in peripheral regions. On the one hand, 'all indicators suggest that information and network societies are metropolitan societies, and that they will continue to be so. Believing otherwise, unless such beliefs are based on solid evidence that contradicts historical patterns, is evidence of faulty brainware' (Abler, 1991, p.42). On the other hand, 'the most vibrant segment in the American economy today is the service sector. Our economy's structural shift to services is not an urban phenomenon; it also affects the rural economy, where public and private services now dwarf agriculture and manufacturing' (Parker et al., 1989, p.4). As with manufacturing, telecommunications may assist the development of services in peripheral regions, but to a limited extent only.

Telecommunications-related services which may potentially develop in peripheral regions can be classified, from a geographical perspective, into two groups, namely regional and national ones. This differentiation relates to the spatial extent of the clientele, and thus to the spatial patterns of information flows, whether primarily within a given region or extending over a whole country. In both cases, the introduction of advanced telecommunications means may potentially involve one of the following three locational patterns: a decentralization of services from major urban centres to peripheral regions; a centralization of services in major urban centres, because of an out-migration of services from peripheries; or a *status quo* of no locational change in services following the introduction of advanced telecommunications means. For regional services one may add another option, namely a growth of services accompanying the development of manufacturing in the peripheries. Let us review these options, beginning with regional services.

Howland (1991) studied several states in the US and could not identify a decentralization of producer services at a regional scale. This was attributed to agglomeration, localization and urbanization economies, especially since 'density of development and a large pool of professional trained workers is lacking' (p.7). However, she was able to identify *regional* growth in producer services, though this may not yet be attributed to industrial developments in those regions. A study of rural counties in the state of Washington, demonstrated that the diffusion of advanced telecommunications means produced a consolidation of regional banking services, or a geographical concentration of producer services, simultaneously with a decentralization of national services (Beyers, 1989).

The Swiss experience in the relocation of banking activities aided by telecommunications is somewhat different, including not only the transfer of service functions, but that of service workers, as well (Erzberger and Sonderegger, 1989). Credit Suisse moved computing facilities from Zurich to outlying workcentres. This move was reasoned by the shortage of computer specialists in the tight labour market in Zurich, the non-central site of the bank headquarters in Zurich causing commuting hardships, and demands for higher life qualities by employees. Interestingly enough, the location of the workcentres was determined, among other considerations, by provincial languages, so that many workers actually returned to their original home regions. This project encountered problems associated with telecommuting, namely limited promotion and transfer options, coupled with a feeling of isolation.

At the national level, low-wage rural and otherwise peripheral regions may become attractive for the location of data entry and processing facilities for large national firms, which produce large volumes of written information, requiring costly manual entry and processing. Prime examples are credit card companies, medical insurance companies, and banks. Documents are mailed to peripheral areas, and the processed information is transmitted electronically to computerized data bases. This type of service activity has developed mainly in rural America. As both Howland (1992) and Beyers (1989) showed, this industry is threatened by lower wages in off-shore locations (which will be highlighted in the next chapter), and by technological improvement, notably in optical scanning and imaging, permitting data entry operations to move back to the major metropolitan centres.

Patterns and problems in selected regions

Coping with the challenge of telecommunications, as well as with resulting development patterns may differ from one country to another, because of differences in the nature of peripheral regions, the proposed plans, government involvement, and the structure of the economy. The following discussion briefly reviews trends in Europe, North America, and developing countries.

Less favoured regions (LFRs) in Europe

The development of telecommunications services and telecommunications-based manufacturing and services in the European, so-called less favoured regions (LFRs) has been recognized as a development target, mainly as of the 1980s. This has been so at both the national level, for example plans for the Northern Region in England, as well as at the EC level, notably the STAR (Special Telecommunication Action for Regional Development) programme. The development problems and the designed policies have further received relatively wide attention in literature (e.g. Lauder, 1990; Goddard and Gillespie, 1986; Robinson, 1984; Hansen *et al.*, 1990; Gillespie, 1987; Pye and Lauder, 1987).

The major less favoured regions in Europe are located in Ireland, Italy (Mezzogiorno), Greece and the UK (Northern Ireland). The problems of LFRs may be divided into two areas, namely those directly related to telecommunications and the more general development problems.

The telecommunications problems amount to gaps between the infrastructure, services and rates available in core areas and those provided in peripheral regions. In the mid-1980s, London's per capita telex penetration rate was five times higher than that of Wales, the least penetrated region. In France, the gap between Paris and Bas Normandie, the respective highest and lowest regions was 3.8 (Gillespie, 1987). Such gaps typified all other measures and services: basic services; waiting lists (e.g., almost six years in Greece in the mid-1980s); service quality (for example, higher call-failure rates in Ireland for calls to and from Dublin); service costs (calling prices are distance-sensitive); data communications; mobile services (Pye and Lauder, 1987; Gillespie, 1987).

Additional problems typify peripheral regions at large. The economic structure of LFRs was based on agriculture and 'low-technology' industries, whereas the production of information technology took place in major urban centres. Also, control was centralized in the major urban centres in the form of company headquarters, and LFRs are normally quite far from such centres (Robinson, 1984). Thus, it was not only that LFRs were typified by a low supply of telecommunications, but they were further characterized by low levels of demand for such services, as well (Lauder, 1990).

The development of telecommunications in LFRs amounted, therefore, to breaking this vicious cycle by developing both supply and demand. It was estimated that if all LFRs in the EC invested 1 per cent of their GDP in telecommunications and information technologies, some 700,000–900,000 new jobs would have been created, at a cost per job that competes favourably with other means of job creation (Hansen et al., 1990). Developments in telecommunications infrastructure, or supply, called, as of the late 1980s, for government-subsidized investments, giving priority to business telecommunications, based on digital technology. As for demand, special assistance has been extended, mainly to manufacturing, but also to services such as tourism. In addition, education plans were proposed in order to upgrade the skill-base of the workforce (Pye and Lauder, 1987; Goddard and Gillespie, 1986; Gillespie, 1987).

Rural America

Rural America shares with the European LFRs several basic disadvantages, notably distances to major metropolitan areas, which in the US and Canada may be even greater than in Europe, the concentration of controlling power in major centres, and the ongoing development of the contemporary information economy in them. However, the American circumstances and experience differ from the European ones in various instances, the most important of which are the available telecommunications infrastructure, the uses of the system, ownership patterns, and the preferred sector for development.

Whereas development efforts in European LFRs have had to cope in many regions with the challenge of introducing basic telephone service to both households and businesses, such service has been available universally in the US for many years. The US enjoys one of the highest levels of telephone penetration, as we noted in Chapter 3. As long as copper-wire technology dominated the telecommunications network, namely until after World War II, local connection to the system of basic telephony was almost universally available. This availability of service was coupled with a system of cross-subsidies, so that remote subscribers were not discriminated by location (Abler, 1991). Thus, an analysis of the approximately 7.8 per cent of US households who did not receive telephone service in 1986 revealed that about two-thirds (64.3 per cent) of them were metropolitan households, only about one-third (32.5 per cent) were rural households with potential access to the system, and merely 3.2 per cent of the households without telephone service, or about 0.25 per cent of all US households had no access to the system because of geographical isolation! The percentage distribution of metropolitan and rural population was similar to the distribution of households without telephone service, namely 64 per cent urban and 36 per cent rural (Parker *et al.*, 1989, p.68).

The gap between rural and urban America is mainly a qualitative one and a contemporary phenomenon, stemming from the development of new telecommunications technologies which require large investments. The rural telephone system in the US is in private hands, like the urban one. However, in addition to the seven large Bell operating companies there are some 1200 telephone companies in rural America. These range from the smallest company, Island Telephone, which serves 34 customers in Frenchboro, an island off coastal Maine, to GTE, the largest one, serving 12,179,000 access lines in 31 states, so that it is as large as a Bell operating company (Sawhney, 1992). It was estimated that at the current rates of replacement of analogue switches with digital ones, the full 'digitization' of rural America will occur in the year 2016! (Parker *et al.*, 1989, p.78). Comparing the bustling global information economy of New York, based on digital and optic-fibre technologies, with this trend reveals the depth of the interregional gaps in the country which enjoys the most extensive information economy in the world.

The universal availability of basic telephone service in rural areas, and the long distances from major urban centres, result in an extensive usage of the existing system. 'In Alaska and northern Canada, native people spend more than three times as much as their urban counterparts on long-distance calls, even though their average income is generally lower than their urban peers' (Parker *et al.*, 1989, pp.34–35). It was also found that the tendency of rural Americans to own and use a wide variety of telecommunications technologies is just as likely as that of urban residents. Furthermore, many rural workers use information in their jobs, so that the human infrastructure for information employment is to a large degree available (LaRose and Mettler, 1989).

The major problem of rural America is, therefore, a financial one, namely the gap between low population densities, on the one hand, and the high investments required for upgrading the existing telecommunications system, on the other. This is obviously not a 'static' problem, namely that a one-time large investment would close the gap, since the continuous innovation and adoption of telecommunications technologies in major urban centres may cause a continuous and maybe growing inter-regional gap. This gap is further stressed by the ownership pattern of the American telephone system. Being in private hands since its inception, there have survived many small independent rural telephone companies, side by side with the Bell system. The investment resources of these companies are restricted, and the assistance provided by the Rural Electrification Administration (REA), the federal agency in charge of developing the rural telecommunications system, is insufficient (Parker *et al.*, 1989).

Advanced telecommunications services may assist farmers, by providing production, marketing, as well as social information (Dillman, 1985). However, as we noted in the previous section, major attempts have been made to transfer regional and national service activities to rural America. The continuous technological gap between cores and peripheries, as well as wage gaps between rural locations and off-shore ones are detrimental for extensive service-based regional development in rural America (Beyers, 1989; Howland, 1991; 1992; Abler, 1991).

Table 5.2 The percentage distribution of population and telephones in urban and rural areas

Country	Year	Urban		Rural	
		Pop.	Tel.	Pop.	Tel.
US*	1986	64.3	64.3	35.7	35.7
Papua New Guinea	1986	12.4	73.9	87.6	26.1
Micronesia	1986	27.4	96.1	72.6	3.9
Industrialized countries	1980	39.8	51.2	60.2	48.8
Developing countries	1980	14.9	70.4	85.1	29.6

* Residential telephones only.

Sources: Parker *et al.*, 1989; Goldschmidt, 1984; Jussawalla and Ogden, 1989.

Less developed countries (LDCs)

The gaps between cores and peripheries in the density of telephone services is much higher in less developed countries (LDCs) than in developed ones (Table 5.2). It was estimated 'that from 60–90 per cent of the population of developing countries generally have access to less than 20 per cent of the available facilities' (Goldschmidt, 1984, pp.182–83). This gap is accompanied by flows of telecommunications traffic centred on the largest urban centres, thus exemplifying the more general dependence on economic activities concentrated in those centres (Saunders *et al.*, 1983, pp.102–3; see also Hudson, 1984). Lesko (1992) found for the province of Tucuman, Argentina, that 73 per cent of the telephone calls involved contacts between the main provincial city and other urban localities in the province. Also, the highest number of calls from the province outwards were addressed to Buenos Aires.

The possible economic benefits of telecommunications developments in the process of regional development in less developed countries are less clear, since this aspect has rarely been studied from an economic perspective. It has been argued, however, that technical progress in telecommunications 'even if available to all regions, may widen rather than narrow the core–periphery gap' (Clapp and Richardson, 1984, p.242). Investing in core areas would widen the social gaps between classes as far as the household sector is concerned (Leff, 1983). For the business sector the gap would also widen, since such investments would imply several negative effects, such as the migration of skilled labour to the core; lower wages in the periphery; and a relatively larger production of goods and services in the core (Clapp and Richardson, 1984).

The role of telecommunications for regional development in developing countries cannot be restricted, however, to economic development, *per se*. Telecommunications, notably through satellite systems, maybe of considerable importance for social objectives, such as the delivery of education and health services (the so-called 'telemedicine'), the transmission of agricultural information and emergency services, as well as permitting social residential calls. Various satellite-based telecommunications systems have been in operation in many countries (see Mowlana and Wilson, 1990; Mayo *et al.*, 1992). Cellular telephony is another technology which may prove useful for regional development in LDCs, since it may require smaller investments in the near future, and since it involves lower safeguard and maintenance costs (*The Economist*, 1992).

Conclusion

If we noted rather conservative spatial patterns in urban areas exposed to contemporary telecommunications technologies, then this is even more so as far as peripheral regions and regional development are concerned. Thus, the interplay between cores and peripheries concerning telecommunications-based economic development is complex, and it may well turn out that cores benefit more than the peripheries as a result of regional development efforts. Still, however, advanced telecommunications facilities have to constitute an integral component of regional development efforts, given the crucial role of telecommunications in contemporary economic activities, as well as its social and cultural importance (see Lesser and Hall, 1987). More vulnerable, and thus more questionable, are efforts to develop telecommunications-based data processing activities. These may require careful technology assessments in order to assure a reasonable longevity for them.

Chapter 6
National Differences in Telecommunications

The objective of this chapter is not to provide a systematic, country-by-country review of the structure and organization of the telephone system, though a variety of country-specific structures and policies will be referred to. A list of works focusing on a single nation or region may be found elsewhere (Brunn and Leinbach, 1991, p.xix). Rather it is aimed here to concentrate on three aspects in which countries differ from each other, and to highlight them both conceptually and comparatively, drawing on the contemporary experience of a variety of countries. These aspects are: the density of the telephone system; national economic development and its relationship with telecommunications; and the recent transitions in the organizational-economic structure of telecommunications systems.

The global distribution of telephones

One may view the global distribution of telephones from two perspectives, namely the absolute number of telephones in countries and continents, and their relative distribution among countries by their population size. The two following subsections discuss these perspectives in this order.

The big gap

The geographical distribution of telephones world-wide is

extremely unequal, presenting a clear pattern of cores and peripheries, similarly to other aspects of economic and social infrastructure and welfare. Back in 1956, one half of the world's telephones were in the US, declining to approximately one-third in the late 1980s (Toffler, 1990). In the mid-1980s, 90 per cent of the world's telephones were concentrated in just 15 per cent of the country-members in the International Telecommunications Union (ITU), and only less than 2 per cent of the United Nations aid was allocated to efforts to close this gap (Daniels, 1985, pp.26–27). In 1990, 121,000 villages in Africa, or about 80 per cent of the 151,000 villages in the continent, had no telephone service, though the African population is 70 per cent rural (Hudson, 1991).

Maybe more than for other elements of development, it is tempting to assess the telecommunications gap between cores and peripheries, from the perspective of an 'advantage of backwardness' (Singlemann, 1978, pp.109–13). In other words, one could think that developing countries may be able to enjoy cheaper and rapidly-installed telecommunications facilities, based on satellite and cellular technologies. Using these technologies they could have avoided the expensive and less efficient copper wires and cables, which still serve as the backbone of the telecommunications system in the industrialized countries. However, this scenario is misleading, since wireless systems could provide universal voice service but only limited advanced data and video services. Thus, while the quantitative gap between cores and peripheries at the global scale may theoretically become smaller, the qualitative one will only increase, and higher levels of quality, variety and volume of service is what contemporary telecommunications systems are all about.

Teledensities

Teledensity (Jussawalla and Ogden, 1989), refers to the national ratio of telephone lines to population, normally the number of main or access lines to 100 persons (Table 6.1). Sweden stands out as the country with the highest teledensity in the world

Telecommunications & geography

Table 6.1 Teledensities by the number of main line per 100 inhabitants 1990

Country	Teledensity	Country	Teledensity
Sweden	67.25	East Germany	10.98
		Argentina	(10.65)
Switzerland	56.91	Turkey	10.35
Canada	(55.80)		
Denmark	55.46	Costa Rica	9.19
USA	53.34	Saudi Arabia	8.94
Finland	52.05	South Africa	8.93
Iceland	50.05	Hungary	8.66
		Panama	8.48
Norway	48.94	Poland	8.25
West Germany	47.43	Malaysia	8.02
France	47.26	Venezuela	7.61
New Zealand	46.45	Colombia	7.36
Luxemburg	46.41	Oman	6.56
Australia	46.32	Brazil	6.01
Netherlands	45.12	Mexico	5.57
UK	(44.58)	Chile	5.18
Japan	43.24	Syria	4.17
Austria	40.72	Iran	(3.69)
Hong Kong	40.72	Tunisia	3.47
		Algeria	3.05
Greece	37.81	Egypt	3.01
Belgium	37.53	Paraguay	2.48
Singapore	37.13	Peru	2.44
Italy	36.97	Thailand	2.09
Israel	34.00	Morocco	1.36
Cyprus	33.63	Zimbabwe	1.36
Taiwan	30.60	Philippines	1.05
Spain	30.40		
		China	0.87
South Korea	28.32	Pakistan	0.73
Ireland	25.88	India	0.57
Portugal	20.07	Indonesia	0.48
		Tanzania	(0.28)
Kuwait	(15.37)		
Yugoslavia	15.00		
Czechoslovakia	14.23		
Uruguay	12.20		
USSR	(11.23)		

Note: ( ) = Estimated values.
Source: Siemens, 1991, p.19.

(67.25), and it has maintained this seniority for many years. The Swedish ratio is much higher than that of the next country, Switzerland (56.91). The seven leading countries, with ratios over 50, are the two North American countries, Scandinavian countries, and Switzerland. As we will see later in this section, the seven leading countries have organized their telephone systems in a variety of ways, as far as ownership and competition are concerned. Thus, high telephone densities may be achieved under various organizational modes, as long as capital is available, and as long as a high priority is given to the development of telephone systems. The organizational dimension is of importance when service prices become critical to continued economic development, and when the 'thickness' of the system, namely the variety and quality of the services provided, are as important as high teledensities.

The next group of countries consists of those with ratios ranging between 40 to 50. These 11 countries include almost all West European countries, as well as the core countries of the Pacific Rim. This group is followed by eight countries with ratios ranging 30–40, and another group of three countries with ratios over 20. These two groups include Southern Europe and the newly industrialized countries (NICs) in the Pacific Rim and the Middle East. Altogether there are only 29 countries in the world with teledensity values over 20.

As one moves in Europe from Scandinavia to central Europe and on to Southern Europe the teledensity ratios are falling (Figure 6.1). This is somehow contrary to what one would assume, namely that central Europe, and especially the three largest countries (UK, West Germany, and France) would lead with lower values presented by the northern and southern countries. However, an early emphasis on universal service in Scandinavian countries, has brought about a north to south pattern of declining values for Europe.

Moving to the next group one realizes a considerable gap between the lowest value in the previous group (Portugal: 20.07) and the leading one in this one (Kuwait: 15.37). This gap attests to the more general one between the core countries and the NICs, on the one hand, and second and third world countries, on the other. Regions with ratios higher than one include

134 Telecommunications & geography

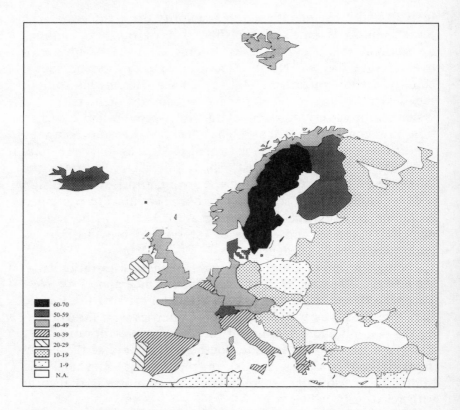

Figure 6.1 European teledensities by the number of main lines per 100 inhabitants, 1990. Data source: Siemens, 1991, p.19.

Eastern Europe, South America, the Middle East, and North Africa. The heavily populated Asian countries are typified by values lower than one, which is also the case for most black African countries.

National economic development and telecommunications

Ranging values of teledensity may be related to measures of national economic development. Such relationships may serve as guidelines for the role of telecommunications in the process of national economic development.

The correlation between measures of development and teledensities

A popular and seemingly straightforward analysis of the relationship between teledensities and economic development would be to construct multinational correlations between national teledensity values, on the one hand, and per capita gross national product (GNP), per capita gross domestic product (GDP), or per capita income, on the other (Cronin *et al.*, 1991). However, such analyses should be carefully interpreted (Saunders *et al.*, 1983, pp.82–84). First, the problem of causation may arise. Does economic wealth 'cause' the development of the telecommunications system or is it vice versa? Second, is one single equation sufficient for the exploration of the complex relationships between the two aspects? These two questions apply mainly to policy decision-making, where one may ask whether a country should accentuate efforts to develop its telecommunications infrastructure merely because it is located above the regression curve and vice versa? Third, is it worthwhile to group together in the same analysis countries with highly sophisticated telecommunications systems with countries whose basic services are still sparse? Beside these limitations correlation analyses may still be valuable for various reasons. First, as we shall see, the causation for the relationship between teledensities and economic development may be explored, leading to some interesting observations. Second, and at a descriptive level, correlation analyses provide some quantitative illustration for the relationship between telecommunications and economic development.

An early correlation analysis was provided by Jipp (1963), and the most extensive ones were performed by Hardy (International Telecommunications Union, 1983; Hardy, 1980). Hardy's study of the 1950s and 1960s covered 15 industrialized countries and 37 developing ones over 15 years. He found that economic development *both* 'causes' or 'leads' telephony developments and vice versa. However, in the developed countries the contribution of telecommunications to economic development was somewhat weaker than in LDCs. This was explained by the larger marginal utility of additional telephones when the system is still small. It was also found that residential telephones led economic development in LDCs. Quantitatively, an increase of 1 per cent in the number of telephones per 100 population between 1950 and 1955 in all 52 countries in the study yielded a rise of about 3 per cent in per capita income between 1955 and 1962. The poorer the country the larger the contribution of telephone developments to the rise of incomes. The importance of residential telephones and the importance of telephones for poorer countries may seem quite contrary to economic thinking, but these observations attest to the universal contribution of telecommunications and to the two-way relationship it has with economic development.

In his study of the 1960s, Hardy (1980) used data from 45 countries for the period 1960–73 with time-lagged offsets of one year. Once again the relationship between GDP and the number of telephones per capita ran in both directions. Also, for this period again, the magnitude of the effect of telecommunications was inversely related to prior levels of teledensities. In other words, a higher economic effect of telecommunications developments was found in LDCs when compared to industrialized countries. Similar results, namely that telecommunications enhances economic development and vice versa, were obtained in a longitudinal analysis of the US, 1958–88, studying investments in telecommunications as against industrial outputs and GNP values (Cronin *et al*, 1991).

A correlation between GNP values for 1989 and teledensity ratios for 1 January 1990 for 43 countries is presented in Figure 6.2. The correlation coefficient is 0.90. The slope of the regression line being higher than 45 degrees indicates that teledensities increase slightly faster than GNP values. It may well be seen that the relationship may be divided into three groups, following the structure presented in Table 6.1 for teledensities. First is the group of industrialized countries enjoying high values of both GNP and teledensity. This group contains, from top to bottom, Scandinavian and North American countries, followed by

National differences in telecommunications   137

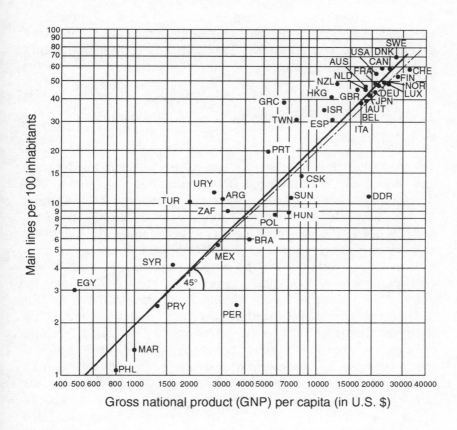

Figure 6.2 Regression between per capita GNP and teledensities for selected countries, 1990. Source: Siemens, 1991, p.21.

Western Europe and Japan, and ending with Mediterranean and Pacific Rim countries. The second group, at the centre of the curve, presents a weaker fit between the two variables and includes mainly South American and Eastern European countries. The last group, at the bottom of the regression line, displays an even weaker relationship, and consists mainly of Middle Eastern and South American countries. Generally then, while it seems that teledensity values go hand in hand with those of GNP the relationship becomes weaker with declining levels of both GNP and teledensity. These trends may suggest that developed countries enjoy and encourage the development of telecommunications, whereas LDCs may put emphasis on other development needs, given the low levels of capital availability and development at large.

The role of telecommunications in national economic development

The relatively low levels of relationship between teledensities and economic wealth for LDCs, and the vast and varied development needs of LDCs, have brought about several views on the desirable share of telecommunications in the development process. These views have emerged despite the relatively high returns on investments in telecommunications infrastructure in LDCs, ranging on the average between 18 and 27 per cent (Saunders *et al.*, 1983, pp.13–14). These views may be grouped into three opinions (Saunders *et al.*, 1983, pp.16–18).

The first view contends that investments in telecommunications should be held at levels below market demands, because of the reasons mentioned above, as well as for political grounds, since telecommunications may potentially be used to bring about political and social unrest. Telecommunications may also enhance migration to urban areas, since it may serve better the urban, notably élite segments of society (see also Leff, 1983). Another perspective calls for telecommunications developments to respond to market demands, which may be restricted by capital availability, as well as by government regulation aiming at the provision of wide public access to basic services. A third view sees telecommunications as a tool for the achievement of widely defined development objectives. Proponents of this view would, therefore, favour aggressive efforts directed at telecommunications infrastructure in order to assist the enhancement of economic, educational, service, and social goals.

Growth in telecommunications systems may assist and enhance several sectors and needs in LDCs. First, it may provide better and wider service to all the previously existing household subscribers, so that benefits from telecommunications investments may increase exponentially. Such wide connections have not merely a social value since they may also enhance more general service provisions, notably in rural areas. They may further assist efforts for national integration (Leff, 1983; Saunders *et al.*, 1983, p.16).

Second, external economies apply not only to households but also to economic firms and organizations in a variety of sectors, ranging from agriculture and industry to commerce, transport and services. With the aid of an expanding telecommunications system they may coordinate their functions better and more efficiently, as well as reach wider markets. Such developments may also imply the creation of new employment opportunities (Leff, 1983; Saunders *et al.*, 1983, p.15).

Third, telecommunications may increase the volume and flows of information in LDCs. This last aspect is of extreme importance when information is in low supply and its flow restricted. Enhanced telecommunications facilities may reduce uncertainty, and it may further permit researchers to contact information bases in the large cities and in foreign countries. Wider, more rapid and more intensive information flows may require not only a well-developed telecommunications infrastructure, but flexible or non-existent regulation on information flows, as well (Pool, 1990, pp.172; 203–4; Leff, 1983).

From 'natural monopoly' to full competition

The organization and ownership of telephone services has probably turned into the most important and 'hottest' aspect of the contemporary telecommunications industry. Major problems are questions such as: should telecommunications services be provided as a governmental public utility service or should they rather function as a business utility service, privately or publicly owned? Should competition be introduced into the provision of telecommunications services? At first sight it seems as if these questions have economic, political and social ramifications, but no geographical ones. However, as the following discussions will attempt to demonstrate, there are at least four ways in which the ownership and organization of telecommunications services have geographical aspects and ramifications.

First, and 'traditionally', the ownership and structure of telecommunications services are determined on a national basis, so that differences in organizational patterns actually amount to variations among nations. Second, and despite the previous point, contemporary conditions in the telecommunications business do not let countries enjoy full autonomy in structuring and restructuring their telecommunications services, though formally they are still autonomous. Once telecommunications services have become a technologically dynamic industry with rapidly emerging and diffusing innovations, and once the provision of such services has to be fast, efficient and profitable, foreign transitions and resulting imitations, cannot be avoided (see also Robinson, 1991).

Third, the impacts of telecommunications services on service industries and manufacturing at large, may create new global maps of the importance and specializations of countries and cities, based, among other things, on differences in the organization and ownership of telecommunications services. Fourth, once various options for the structure and ownership of telecommunications utility companies become feasible, companies' capital may be invested in foreign telecommunications companies, thus creating new interdependencies. The following subsections will, therefore, be devoted to the following aspects: the various ownership patterns; the emergence of this variety; and the growth and geographical implications of these organizational transitions in telecommunications services.

Organizational and ownership patterns for telecommunications services

Two characteristics typified telecommunications services until the mid-1980s: they were considered 'natural monopoly', and with only two permanent exceptions (the US and Canada), they were operated by national governments as an integral part of PTTs. The term natural monopoly relates to the unique character of the supply of utilities at large, including gas, water, electricity, and telephone services. These utilities require connections by pipelines and cables at each served point, and competition among various networks, could amount to physical, and in some cases also financial chaos. The role of economies of scale is, therefore, crucial for utilities which have to reach every building. However, whereas various utilities were often supplied by a private monopolistic company, telephone services were considered, in almost all countries, a government service. They were somehow similar to public transportation, where route-networks, stops and terminals have to be structured and maintained, competition is usually absent, and the service provider charges users for each ride. Like in most public transportation services, profitability was not required in telephone services, and when capital surpluses accumulated, they were transferred to the general PTT budget or to the national budget.

The technological transformations of the telecommunications industry permitted several changes in the double-character of telephone services, namely its being a natural monopoly, and its being offered as a governmental service. First, it has become possible for competing companies to offer telecommunications services, at least long-distance ones. Second, newly invented technologies permitted two new options of service provision, namely using the existing system for new services (mainly fax and data transmission), and offering services without cable-networking (mainly cellular telephones). Third, telecommunications has become a service that requires heavy investments on the supply side, and it has further turned into a crucial input on the demand side, whether in form of producer services, manufacturing or households. Fourth, and at a later stage, contemporary telecommunications has become by its very nature a global industry, so that foreign influences and investments were called for.

These new dimensions of telecommunications have brought about, within less than a decade, a structural change in the ownership and organization of telephone services, so that several modes have now become possible (Table 6.2). These options run from the most restrictive and traditional form, namely PTTs, through various other forms of governmental ownership, to private ownership and competition. The technological sophistication of contemporary telecommunications systems and the

Table 6.2 Various ownership options for telecommunications services

Option	Examples	References
1. PTT	Belgium[1]; Singapore[1]	*The Economist*, 1991; Corey, 1991a; 1991b; Heng and Low, 1990
2. Telecommunications administration	Germany	Schmidt, 1991
3. Government-owned company	France; Senegal; Zambia	Akwule, 1991; Hudson, 1991; *The Economist*, 1991; Saunders *et al.*, 1983
4. Shared ownership by government and private capital	Israel; Hong Kong; Portugal; Nigeria; Argentina	Ure, 1989; Case and Ferreira, 1990; *Bezek*, 1991; Lesko, 1989
5. Private domestic ownership without competition	Canada	Staple, 1991
6. Private foreign ownership without competition	Mexico; Hong Kong[2]	Staple, 1991; *The Economist*, 1991; Ure, 1989
7. Competing private, domestic companies	USA[2]; UK[2]; Japan[2]; Australia[2]; New Zealand	Akhavan-Majid, 1990; Glynn, 1992; *The Economist*, 1991; Langdale, 1991a; Dordick, 1990; Phillips, 1991
8. Competing private, domestic and foreign companies	–	Staple, 1991

[1] Change was expected in 1992.
[2] For long-distance and international calls only.

complex interrelationships among service agencies may sometimes introduce competition even if unwanted, through various kinds of bypassing, notably in international telecommunications. Thus, lines may be leased and then resold on a call-by-call basis, with several calls transmitted simultaneously through a single line, or calls may be placed to a domestic number which is connected to a computerized exchange in another country offering cheaper rates.

The changes in ownership patterns are *not* necessarily related to a nation's economic development. France and Germany prefer more restrictive and public ownership modes, whereas Argentina and Mexico permitted private or foreign ownership respectively. It may further be noted that Western European nations, with the exception of the UK, prefer more conservative forms of organization, while developed countries in other parts of the world demonstrate more openness for private ownership and competition. The reorganization of telecommunications systems is still in process in many countries, so that it is difficult to present a definite international comparison at this point in time.

The various options in Table 6.2 reflect three aspects simultaneously: economic liberalization; functional deregulation; and power distribution. Economic liberalization on its part relates to three dimensions of ownership and organization of telecommunications services, namely the possible permission of competition, private ownership, and diversification of capital sources. In these three areas of economic liberalization, as well as in the other two aspects, countries differ from each other. Functional deregulation, refers to the telecommunications services which companies may operate, and under which conditions. This applies to both the types of offered services (voice telephony, fax, data transmission, video services, cellular telephony), as well as to their geographical range (local, long-distance, international). Liberalization and deregulation usually go hand in hand, though a high level of change in one of the two does not automatically call for such a level in the other. Power distribution is also interwoven with liberalization and deregulation. The power of telecommunications provision may rest in government or in business. Telecommunications may thus be considered a public service or it could rather turn into a source of profitability.

The emergence of organizational and ownership patterns

The various ownership patterns presented in Table 6.2 emerged during the 1980s and they continue to undergo transformation in the 1990s. It is important to focus on leading industrialized countries in order to find out which countries led the change, why it affected the telecommunications industry on a global scale, and how did the various ownership patterns shape up.

The transformation in the ownership patterns of telecommunications services was led by the US, and this seems peculiar at a first glance, when it is recalled that the US and Canada were historically different in that telecommunications services in these two countries were privately owned since their inception. Thus, any change in the American private system should not have affected other countries which traditionally have preferred a completely different ownership-path, namely PTTs. The global shaking of telecommunications systems ignited by changes in the US is related to the increasing role of telecommunications in the emerging service economies, notably their international component, which on their part were related to the vast technological breakthroughs in telecommunications. Organizational changes in telecommunications in the US, the country which has led the evolution of service economies since the 1960s (Kellerman, 1985), have meant that the US has become more competitive in its service economy, a challenge which could not be left unmet by other leading countries. Thus, the transformations of telecommunications organizational patterns are, at least in part, related to the evolution of the global economy. It is clear, therefore, that the US leadership in telecommunications innovations has not been restricted to technology, but it applies to ownership and organization patterns, as well.

Invented in the US, the telephone received a similar attitude as its predecessor, the telegraph, namely that Congress did not show an interest in buying the patent, so that it was left for private ownership and development. The development of the telephone system by the Bell Company, or AT&T, was aided by the expansion of the US in the late nineteenth century, a factor which was well-perceived by Theodore Vail who headed the company at that time. When the Bell patent expired in 1893, the company was able to compete efficiently, especially once it adopted direct-dialling in the 1920s (Abler, 1991; Dordick, 1990) (see also Chapter 2). It still remains to be studied how the modern economic development of the US was aided by the universal availability of telephone connection, rather than businesses and households being at the mercy of government bureaucracies as in many other countries.

In 1934 the Federal Communications Commission (FCC) was established and charged with the mandate to monitor and control the US telecommunications system. The deregulation of the US telecommunications system has gone hand in hand with technological developments which were challenged by customers and entrepreneurs. The 1950s witnessed the deregulation of private microwave transmission, known as the *Above 890* decision, followed by the permission given in 1969 to MCI to establish commercial microwave transmission lines (Langdale, 1983; Phillips, 1991). Attempts by the federal governments since the 1940s to challenge the monopoly of AT&T on the grounds of excessive customer charges failed. The emergence of computers in the 1960s was accompanied by demands for computer communications presented by IBM. These were not met by AT&T, and eventually brought about, in 1982, the divestiture of AT&T into seven regional Bell companies ('the baby Bells' or RBOCs), in charge of local communications, and a national AT&T dealing with long-distance and international calls in a competitive market. AT&T was permitted in return to enter the computer business (Toffler, 1990).

Two countries were fast in coping with the American challenge of rapidly declining prices, coupled with higher qualities and a larger variety of telecommunications services. The UK and Japan had their systems privatized and opened to competition in 1984–85. There were beginnings for this process in the early 1980s, at least in the UK, and there were several factors behind it. However, the very privatization and deregulation processes in these two countries, as well as their pace, were influenced, at least in part, by deregulation in the US, thus aiming at safeguarding their competitiveness in global financial markets, operating in London and Tokyo respectively (see Chapter 4).

Another reason for the UK move was the British desire to serve as the international telecommunications hub for Western Europe (Langdale, 1989a).

The British romance with privatization and deregulation may be traced back to July 21, 1980, when the then Secretary of State for Industry, Sir Keith Joseph, informed Parliament:

> We are going to allow people more freedom to use British Telecommunications' circuits to offer services to third parties which are not currently provided by British Telecommunications, for example in the data processing field. I expect this change to lead to a significant growth in information, data transmission, educational and entertainment services provided over telephone circuits and to the emergence of new business. I have also decided to commission an independent economic assessment of the implications of allowing complete liberalization for what are commonly referred to as value added network services (Beesley, 1992, p.223).

This policy declaration led a year later to the British Telecommunications Bill, which separated telecommunications from postal services, and which gave the Secretary of State the powers of licensing independent operators, which, on their part, could either use (lease) lines from British Telecom (BT) or could provide their own networks. In 1984 BT was privatized and in 1985 a second PTO (Public Telecommunications Operator), Mercury, started operation, thus creating a duopoly in the British system. In the early 1990s the duopoly was formally broken.

In Japan, the Diet (Parliament) passed two laws in 1984, which privatized, as of 1985, Nippon Telegraph and Telephone (NTT), and deregulated the provision of telecommunications services, so that three companies started competition (Akhavan-Majid, 1990). These steps resulted from pressures by the US as well as by Japanese big business for liberalized markets. The Japanese system attempts to assure national economic growth and the provision of social needs, side by side with profitability of the carriers. Thus, external subsidies, rather than cross-subsidies are provided for desired yet unprofitable services (Glynn, 1992).

The French and German approaches to change in the telecommunications system have been completely different than the

American, British, and Japanese ones. In France, the development of the telecommunications system until the early 1970s lagged considerably behind that of other industrialized countries. France was able to enjoy an annual economic growth rate of 6 per cent from 1955 to 1970, concentrated in the primary and secondary sectors, and in specific tertiary activities concentrated in Paris (Saunders *et al.*, 1983, pp.87–88). In the mid-1970s the French government realized that further economic growth will occur in the tertiary sector and that regional dispersion of industrial production is desired, and that both trends are telecommunications-dependent (Nora and Minc, 1980). Thus, a major, and probably unprecedented, governmental effort was directed towards closing the telecommunications gap and towards the modernization of the system. Only as of the beginning of 1991, the PTT structure was reformed, so that France Telecom, a state-owned company was established (Staple, 1991). However, an advanced telecommunications system and an advanced service economy functioning in a highly regulated environment may call for low flexibility in the provision of sophisticated services (*The Economist*, 1991).

Germany too had its telecommunications system anchored in its PTT (Deutsches Bundespost, DBP) until 1989, when a separate monopolistic telecommunications administration was established (Schmidt, 1991). The high level of services and the special concessions granted to financial institutions delayed the pressures for reform presented by high-tech industries as well as by political parties until after liberalization was introduced in the UK. The new system has left only 10 per cent of the market open to competition, mainly leased lines and cellular telephony. Though the new organizational system has turned out to be complex, further reform has been slowed down by the integration of the old and lagging East German system into the German Telekom (Schmidt, 1991).

One cannot conclude this very partial review of major changes and trends in ownership patterns without mentioning the Mexican experience. In 1990 Telemex was privatized by selling the company to domestic and foreign investors (American and French). The deal called for generous tax conditions, assuming an annual increase of 12 per cent in the number of exchange lines (*The Economist*, 1991; Staple, 1991).

Growth implications of the organizational transitions in telecommunications services

Organizational and ownership transitions in telecommunications services may have wide implications for the growth of telecommunications systems, in both the expansion and modernization of the infrastructure, as well as in the production of larger traffic volumes and resulting higher revenues.

As far as infrastructure is concerned, two 'bottle necks' were typical of PTT systems. One was the inability to rapidly expand the system, or supply, so that increasing demands for telephone lines created long waiting lines. The other problem was a shortage in capital required for the modernization of the infrastructure via the installation of digital exchanges. Even a modest organizational transition, namely from PTT to a government-owned company with its own labour hiring and compensation policies, as well as its own budget and business policy, may bring about major changes in the expansion and modernization of the system. The establishment of *Bezek*, the Israeli telecommunications company in 1984 as a state company, and as a partially privatized company as of 1990, brought about extremely fast relief in the waiting line problem (Figure 6.3). Two years after the establishment of the company the number of installations exceeded the number of waiting applications despite a continued growth in the number of applications, and as of 1991 most applications are fulfilled within days or weeks.

Digitization has also been aided by ownership changes. The share of digital lines in Israel in 1992 was 60 per cent, growing at a rate of 5-10 per cent annually. In Hong Kong, a city-country with a similar teledensity as Israel, albeit smaller in territorial size, digitization completion is expected by 1993 (Ure, 1989). The geographical distribution of digital lines may not necessarily discriminate against peripheral regions. Thus, the service areas with highest and lowest rates of digitization in Israel are both located in the urban field of Tel-Aviv (*Bezek*, 1991), whereas the previous analogue technology presented wide regional gaps (Salomon and Razin, 1988).

The introduction of competition may enhance both traffic growth rates and real revenue growth rates (Figure 6.4). The

National differences in telecommunications   149

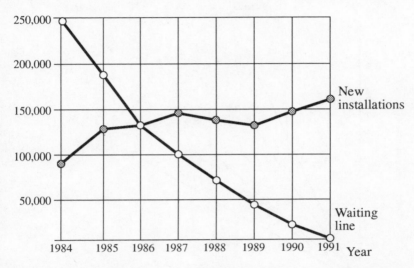

Figure 6.3 Waiting applications for telephone lines and new installations in Israel, 1984–91. Source: *Ma'ariv*, 1992c.

relationship between these two measures of growth was better between 1983–88 for Japan, the UK, and the US, all of which introduced competition, when compared with Germany and France which did not. Such a relationship may shed some light on the expansion of traffic, or the wider impacts of telecommunications, as against revenues, or the stricter financial measures results of the system (Glynn, 1992). The difference between the countries which introduced competition (US, UK, and Japan), on the one hand, and those which have not (France and Germany), on the other, reflects a difference in approach to service provision. Whereas governmental monopolies may tend to look at telecommunications systems from their supply side, namely through the provision of infrastructure and services, competitive regimes wish to achieve, foremost, optimal uses of the network through enhanced levels of demand (see Robinson, 1991).

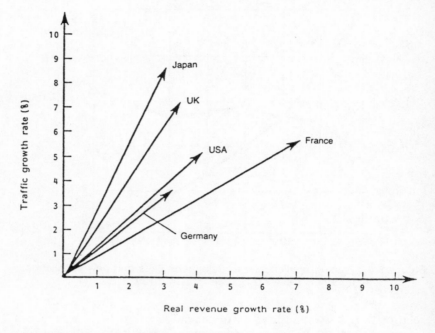

Figure 6.4 International comparisons of telecommunications traffic and real revenue growth, 1983–88. Source: Glynn, 1992.

Geographical implications of changing ownership patterns

Privatization and deregulation may influence changing geographical balances between cities and regions at both the domestic and regional levels. Langdale (1991a) warned of a possibly changing power distribution in Australia once its telecommunications company, Telecom, will compete with new carriers. He claimed that Sydney and Melbourne will enjoy the services of several carriers, and at lower prices, whereas other regions will not be able to enjoy these benefits. Similar phenomena emerged in the UK and the US following the introduction of competition. Thus, the British Mercury preferred to invest in the construction of its network in the Southeast and selected other major cities, and calling rates in the US were lowered for routes with high levels of demand, while others were increased.

At the international level, Keen (1988, p.277) argued that:

> No city can be a really major business location in the world network without being a major money center; that requires electronic international banking and security trading. It must, too, invite rather than push away multinational corporations that are using computers and communications to evolve a federated structure; that requires a national policy of liberalization of telecommunications.

According to Keen (1988, p.278), this provides an explanation for the relative weakness of Frankfurt and Paris in global capital markets compared with London, New York, and Tokyo. Though the German mark is a major international currency, the restrictive German telecommunications policies, coupled with the preferred time-zone, bring about a trade in foreign exchange in London that is ten times as much as Germany's. By the same token, the restrictive French telecommunications policies have prevented the Paris airports from turning into major hubs in the world trade network.

Conclusion

There are marked differences in the penetration levels of telephones among nations. Though these differences clearly fit the international gaps in economic development at large, it is less clear how these two measures are interrelated.

Another variation among nations concerns the structure and ownership organization of telecommunications services. When telephone systems are quantitatively compared with levels of economic development, then these two aspects seem to be unrelated, since high levels of teledensities were achieved by countries adhering to a variety of organizational patterns. However, the emergence of diversified and sophisticated telecommunications means may call for the adoption of new comparative measures, such as call prices, revenues relative to system expansion, the availability of services, and the like. Such measures may show an advantage for countries which have liberalized and deregulated their telecommunications systems.

Since most countries have adopted some change in the organizational structure of telecommunications services, and since the traditional PTT system is almost non-existent, one may conclude that telecommunications systems gradually change from a governmental monopolistic utility to a market dominated business.

Chapter 7
International telecommunications

The availability of directly dialled, high quality, and reasonably priced international telecommunications is no doubt one of the most striking facets of the telecommunications revolution, for both businesses and households. There are several aspects which are more striking in international telecommunications than in domestic telecommunications. First, the barriers for connection are more significant, whether administrative, through the international account settlement agreements, or culturally in the form of language differences. In addition, time differences are notable in intercontinental calling, as are also political barriers and call-price differences between two parties engaged in an electronic information exchange. Second, international telecommunications seems to be more heavily related to other international movements or exchanges, which also encounter various barriers, namely the exchanges of commodities, people and capital. Third, international telecommunications is the only form of telecommunications for which at least basic geographical data are published, either by national telephone companies, by governments, or in special annual publications (Staple, 1991; AT&T's *The World's Telephones*; Siemens, 1991). Though the data are aggregate, showing the annual total number of calls, and/or the number of call minutes to (and sometimes from) other countries, without a breakdown into the calling sectors, this is more than the total lack of geographical data for domestic calls.

The following sections will discuss international telecommunications along the sequence presented in Figure 7.1. Following an elaboration of the importance of international telecommunications, its movement aspects will be highlighted in light of other

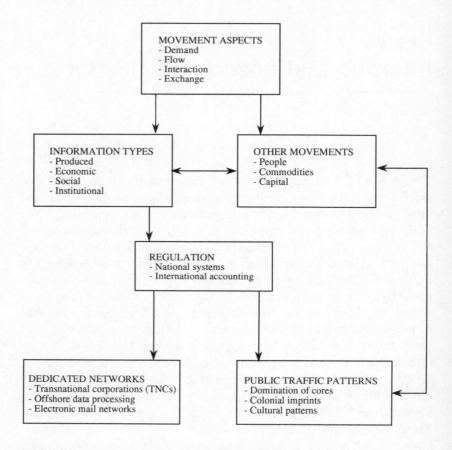

Figure 7.1 Dimensions of international telecommunications.

movements and the several information types. Next, attention will be given to the regulation of international telecommunications, with a special emphasis on the gap between old arrangements, on the one hand, and new technologies and ownership patterns, on the other. The discussion of international networks, which will follow, will be divided into two sections, namely public networks and dedicated ones. The section on public networks will focus on patterns of traffic and called countries, and on factors for these patterns. The section on dedicated networks will present some of the more frequent uses of such systems.

Basic dimensions of international telecommunications

It was merely 11 years after the introduction of the telephone in 1876 when it was implemented for international communications, namely between Paris and Brussels in 1887, and it took only another four years for underwater telephone cables to connect England and France in 1891 (Kern, 1983, p.214; see also Table 2.2). However, international telephony remained secondary to telegraph and later on to telex services until the mid-1960s. It was then, with the introduction of satellites, digital switchboards, and the resulting high-quality, directly-dialled and reasonably-priced international services, that international telecommunications has turned into a crucial element for almost any international activity.

The importance of international telecommunications

International telecommunications has become important from a variety of perspectives, namely the economic, political, social, and geographical ones (Kellerman, 1991b). Economically, telecommunications has proven itself as an indispensable component of the emerging global economy. We have noted in Chapter 4 the evolution of a global urban system in this regard, based on a variety of capital markets. In later sections in this chapter, we will elaborate on the dedicated networks which serve this system, mainly for data transmission, as well as on the traffic patterns in the public telephone system. Later sections will further highlight additional economic dimensions of international telecommunications. Thus, international telecommunications permitted a geographical peripheralization of economic activity similarly to such processes at the domestic level. Telecommunications has also become an integral element in the international exchanges of capital and commodities.

Politically, it was argued that telecommunications may prevent the outbreak of wars (Lyon, 1986, p.580), as well as promote international understanding (Kellerman, 1991b). It was in the attempted coup in the Soviet Union in August 1991 that

156 Telecommunications & geography

Figure 7.2 The role of international telecommunications in the attempted coup in the Soviet Union, August 1991. Source: *The Washington Post*, August 29, 1991.

the open channels of all telecommunications means helped Yeltsin and Gorbachev to win the battle (Figure 7.2; Hoffman, 1991; Henry, 1992). While Yeltsin's struggle in the Russian parliament could be watched over CNN broadcasts by 100,000 Moscovites, as well as by additional thousands in other Soviet cities, Gorbachev was able to listen to Western supportive radio news, and political support was transmitted from Western governments via open fax and telephone lines. This incident may have signalled the fading of national sovereignty over information, as well as the power of the US in the provision of global instantaneous information (Henry, 1992).

From a social perspective, international telecommunications provides a higher quality of life, when social and family contacts can be strongly maintained (Kellerman, 1991b). This is especially noticeable in immigrant countries (Langdale, 1991b). International tourism, both as an economic activity and as a leisure activity, may too be enhanced by telecommunications (Langdale, 1991b).

Geographically, international telecommunications reduces the friction of distance even more considerably than in domestic telecommunications. This permits cooperation in the construction of telecommunications systems over very wide geographical areas. Thus, the tiny island-nations in the Pacific basin have developed the PEACESAT project for joint telecommunications and information services in a mostly maritime area (Lewis and Mukaida, 1991). By the same token the southern African nations have cooperated in the provision of telecommunications services over an extensive continental territory (Dymond, 1987).

Movement aspects of international telecommunications

There are four types of information which we recognized in Chapter 3, namely produced, economic, social, and institutional. In addition to these types of information, three other elements may move internationally (as well as domestically): people, commodities, and capital. The movements of all these elements may be compared along four aspects: geographical aspects of *flow*; economic aspects of *demand*; social ones

Table 7.1 Types and characteristics of international movements

Aspects Characteristics	Geographical Flow	Economic Demand	Social Interaction	Political Exchange
Types				
Information	Variously instant telecommunications (telephone, telex, telegram) or delayed communications (air and surface mail). Serving households and businesses. Difficult to separate transactions by contents.	Data and business information may be stocked and are durable. Personal information may be recorded. Prices are partially regulated.	Direct person to person in telephone communications; is sensitive to common working and leisure hours, and to language barriers.	Technology improvements enhance immediacy and reduce institutional intervention per each transaction. No preparation and 'packing' necessary. Industry still subject to international agreements.
Commodities	Fast and slow delivery (air and surface shipping). Almost completely commercial. Classified by commodity type.	Durable in various degrees and can be stocked. Prices are usually regulated.	Seller to buyer communications; frequently less sensitive to immediate language barriers.	Institutional involvement in almost all transactions, even if cleared of customs and permits. Preparation and packing necessary. Industry subject to international agreements.

contd.

Table 7.1 contd.

Aspects *Characteristics*	Geographical *Flow*	Economic *Demand*	Social *Interaction*	Political *Exchange*
People	Fast and slow traffic (air, boat and maritime travel). Travel for business, personal and immigration purposes. Differentiated by travel purpose.	Travel cannot be stocked, and is not durable, except for immigration. Prices are usually regulated.	Direct person-to-person communications; less sensitive to language barriers than telecommunications.	Institutional involvement in each journey even if visas are unnecessary. Preparation and packing necessary. Industry subject to international agreements.
Capital	Fast and slow flows (via mail or telecommunications). Almost completely commercial. Identified by investment sector or type.	May be stocked and is durable. Transfers, interest rates and taxing are usually regulated. Stock transactions and business management are sensitive to common working hours.	Institutionalized transactions; less sensitive to language barriers, but sensitive to common working hours.	Institutional involvement in most transactions, sometimes by reporting only. Preparation and 'packing' sometimes necessary. Industry subject to international agreements.

Source: Kellerman and Cohen, 1992.

relating to *interaction*; and political ones concerning *exchange* (Figure 7.1; Table 7.1; Kellerman and Cohen, 1992).

One may compare the four movement types to the three possible states of matter. Moving information with the use of modern telephone technology is as flexible as gas; it may change modes, shapes and volume easily and its transfer is fast, indeed virtually instant. The movement of people is similar to liquid. It may change travel modes and it is self-motored to some degree, but it is not instant and requires preparations. The movement of goods is like (and frequently is) solid. Moving is slower and requires handling. The movement of capital used to be closer in its nature to the movement of people and commodities, but has recently become similar to the movement of information, though capital transfers may still be regulated and reported.

The enormous flexibility of information movement over the telephone may be internationally restrained by two major barriers, namely language and time differences, which are less or irrelevant in other forms of international movement. Additional forms of information movement such as telex or mail may overcome these obstacles, but then the immediacy, ease of use and informality are lost.

Each of the four types of movements may be sub-classified by purpose or kind. Moving people, information or capital may be classified as business or personal, while moving commodities is usually classified by types of goods. However, since moving information is so instant and easy, it has practically become impossible to classify it into the four types mentioned earlier, while data for the other types of movement are almost always sub-classified by purpose or kind.

The impossibility to divide information movement into its major sources seems to strengthen the question whether it is related to types of movements of economic nature or whether it is more closely related to social-personal relations. On the one hand, the movements of goods, people and capital all require information exchanges. On the other hand, social-personal telephone conversations are the communication form closest to face-to-face meetings. It is possible, therefore, to argue that the movements of people and information both contain a major social element.

Keeping in mind the enormous flow flexibility of telecommunications, and its high dependency on other movements, we may examine international telecommunications as spatial interaction. In his classical study of American commodity flows, Ullman (1957) put forward three concepts for its analysis: transferability, complementarity, and intervening opportunity. It is difficult to apply the last concept to conditions of instant and often even distance-insensitive flows of information. There are, however, other factors of telecommunications with effects similar to intervening opportunities. For example, time differences and language barriers, or restrictions on direct-dialling, as well as the quality of connections and the prices for various forms of transmissions, may well determine the volume of communications, the use of fax or telephone, the amount of data transferred, and imbalances in the two-way movements of information.

The complementarity of international telecommunications is expressed in two ways. One is in produced information, and is similar to complementarity in the flow of commodities. Information which is unavailable in one country may be produced in another and then transmitted to the first. A second form of complementarity is unique to telecommunications and it refers to the completion of economic, social and institutional contacts *vis-à-vis* telecommunications. From the telecommunications perspective this refers to the dependence of the movement of information on other movements types.

Spatial interaction consists of three components, namely the terminals or the interacting nodes, the channels or transportation routes and means, and the substance of the interaction (exchange of goods, people or ideas). It is important to draw some attention in this regard to a comparison between telecommunications on the one hand, and transportation and international trade on the other. Such a comparison will shed some light on the transferability perspective.

There are differences between transportation and telecommunications which relate to technology, the classification of 'loads' and the differentiations between types of users (businesses and households). Technologically, it was noted elsewhere that the telecommunications industry is more complex than transportation, given the need for hardware and software, and the specialized software, equipment and terminals required for each use (Kellerman, 1984). The load in transportation is usually distinctively separated into people and cargo, both on the vehicles and vessels and in the related statistics. In telecommunications, as we noted, it is difficult to separate the four types of information, or even business and social messages. Finally, in transportation it is easy to separate users by type of vehicle, so that households typically use private cars and rarely own trains or airplanes. In the telecommunications business it is becoming common to find households equipped not only with the traditional voice telephone but with computer and fax 'business-like' telecommunications means as well.

The emergence of international trade has been classically attributed to supply and demand, comparative advantage and a relative lack of barriers. However, the extensive use of international telecommunications for social purposes (which is sometimes blended with business uses), coupled with the enormous ease of transmission, make it difficult to analyse international telecommunications along the lines of international trade. There exists, however, a comparative advantage in international telecommunications, as well, though in a different way. Calling rates from one country may be different than calling back, while in international trade shipping rates are usually equal in both directions. Also, the telephones at both ends of an international connection might be similar but the operating organizations different (governmental, public, private or competitive).

The regulation of international telecommunications

Among the features of international telecommunications, relative to other forms of international movement, is the 'transparency' of its regulation to its users. Thus, whereas one's crossing of a border or shipping goods to a foreign country requires documentation and preparation of each transfer, placing calls or the transmission of data can be performed instantly in most cases. Though there is much regulation behind the scenes of each communications activity, the user may feel it only in the charges paid for communications. The regulation of international telecommunications is double, consisting of domestic and international components. Domestic regulation was the subject of the previous chapter, so that this section will be devoted to the international regulation of telecommunications, focusing on the International Telecommunication Union (ITU), and on the international accounting for telecommunications.

The International Telecommunication Union (ITU)

The history of the ITU was recently sketched by Codding (1991), who identified the ITU as the oldest international organization, established in 1865 as the International Telegraph Union. Located in Geneva, Switzerland, its history may be divided into four periods: (1) until 1934, when the International Telegraph Union merged with an equivalent organization, established in 1903 for international cooperation in radio issues (including radio-telephone), in order to establish the ITU. The merger reflected the similarity of problems in the two fields as well as the reality of national PTTs dealing with both fields; (2) 1934–45, a period during which radio technology improved, but the ITU showed little activity; (3) 1947–65, years which were dominated by reorganization and emphasis on radio frequency problems; (4) from 1965 to the present, characterized by coping with computerized telecommunications, satellites, and problems of developing countries which have become the majority in the organization membership.

Traditionally, then, the ITU allocated radio frequencies and regulated the accounting among national telecommunications agencies, using conferences and consultations as its main operational channels. As of the late 1950s, the ITU has been increasingly involved in development efforts in LDCs (Ono, 1990). In 1959 it was decided that development assistance would be provided through participation in United Nations plans. As of 1982, the ITU allocated funds of its own for development purposes, accentuating training, long-term planning, cooperation with other international agencies, and the dissemination of

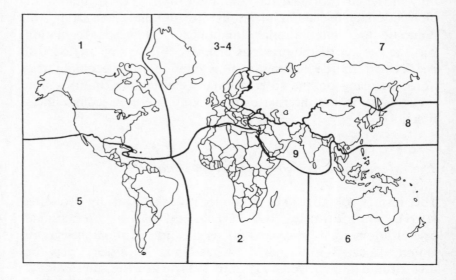

Figure 7.3 World zones for country first dial digit. Source: Abbatiello and Sarch, 1987, p.81.

information. In 1984 and 1989 the ITU established its own development agencies, the Center for Telecommunications Development (CTD), and the Telecommunications Development Bureau (TDB), respectively. Given its long history and authority, various calls have been made for the ITU to serve as leader in the coordination and consultation of development efforts (Hudson, 1991; Ono, 1990).

Another responsibility of the ITU is the division of the world into international calling zones. The first dialling digits 1–9 were allocated to world regions (Figure 7.3). This division reflects political realities of the 1960s, so that the Soviet Union received its own dialling zone, like the US and Canada together, despite its low teledensity. Also, the large number of countries in Europe with high teledensities required the allocation of two first digits for the continent. The exact country code was determined by country size, ranging from one to three digits. This system may undergo some changes in 1993, as the countries of the ex-Soviet Union would like to move from zone 7 (Soviet Union) to zones 3 and 4 (Europe) (*The Economist*, 1992).

International accounting for telecommunications

The purpose of the international accounting system for telecommunications is to settle the costs of provision of a service that originated in one country and terminates in another. The transmission of calls from any telephone subscriber in one country until that country's international gateways, whether through cables, radio stations or satellite antennas, is the sole responsibility of the domestic carrier. International cables usually constitute joint enterprises, and satellite services are leased or bought from INTELSAT or other companies. However, the domestic company serving the call sender has to pay for the transmission of the call from the gateway of the receiving country to the called subscriber. This portion of international calls is the object of the international telecommunications settlement arrangements.

Until 1944 one could recognize three systems of arrangements, representing three powers, namely the British Empire and later the Commonwealth dominated by Cable & Wireless, the West European nations functioning through their PTTs, and the US where most of the service was provided by AT&T, and regulated, as of 1934, by FCC. The British attempted to divert as much traffic as possible through London, the hub of their global system. The Europeans negotiated fixed bilateral agreements, while the Americans, notably following the establishment of RCA as the radio service provision company after World War I, attempted to minimize the costly use of the British system (Ergas and Paterson, 1991).

The universal system which was developed in 1944, and which generally prevails until now, is based on a distance-based fixed charge, usually per minute call, which is generally equally divided between the sending and receiving parties. The charge may be defined in US dollars or in SDRs (Special Drawing Rights; a currency basket established by the International Monetary Fund). This arrangement was successfully negotiated by AT&T and FCC at the time for three reasons: the changing geopolitical balance from a British dominance to an American one; the nationalization of international telecommunications services in most Commonwealth countries; and the modifications in radio-telephony and the later construction of maritime cables required the use of the shortest transmission routes (Ergas and Paterson, 1991).

The settlement proved itself a convenient one until the telecommunications revolution of the 1970s. It was simple, universal, permitted a geographically-efficient service as well as investments in infrastructure, and above all, it assumed equal players in the telecommunications game (Ergas and Paterson, 1991). However, technological improvements and later on the introduction of competition have lowered call tariffs in some countries, notably in the US, so that payment deficits have emerged.

The annual payment deficit of the US reached $2.4 billion in 1989, growing at an annual rate of 22.6 per cent (Stanley, 1991). The US was in surplus with only 17 of its 188 correspondent countries in 1987, with a total net outflow of calls equivalent to almost one-third of the US international traffic (Cheong and Mullins, 1991). The deficit was attributed to several factors:

> (1) significant disparities in prices and pricing policies for service in the United States as compared to service in other countries; (2) an international settlement procedure that is unresponsive to changing demand and supply conditions in the industry, which impedes remedial steps; (3) differences in regulatory environments, goals, and initiatives in the United States and other countries; and (4) differences in incentives and market conditions in which entities provide international communications service in different countries. Other factors, such as per capita income differences and fluctuating exchange rates, also contribute to the deficit (Stanley, 1991, p.412).

The relative importance of each of these factors is still questionable. It was noted that the UK and Japan also adopted a competitive system, but they did not suffer deficits in the 1980s, whereas Australia and Canada which did not do so in the 1980s did present deficits. The US deficit was, thus, attributed to prolonged surpluses in demand (Cheong and Mullins, 1991), maybe because of the American tendency to make more use of the telephone. Also, statistical regression analyses using US and Australian payment balance data with other countries as dependent variables and GDP of those countries as independent ones proved useful (R-square = .55) (Ergas and Paterson, 1991; Cheong and Mullins, 1991). Thus, while countries with an annual per capita GDP below $5,000 accounted in 1987 for only 18 per cent of Australia's two-way traffic, they received 55 per cent of outpayments. Such countries generated 39 per cent of AT&T's traffic, but received 62 per cent of the payments (Ergas and Paterson, 1991). However, there are wealthy countries with high-calling tariffs, such as Germany, which ranks very high on the US list of deficit countries.

The existing settlement system does not permit full use of the telecommunications infrastructure, and it punishes the cheaper, innovative and more efficient side. Furthermore, it encourages the development of high-demand routes at the expense of more modest ones. On the other hand, it controls the system in developing countries, so that surplus–demand for international telecommunications is regulated by high prices, and international telecommunications also becomes a source of foreign exchange for them (Ergas and Paterson, 1991). Current pressures by countries which introduced competition will probably bring about some change which will reflect declining costs as well as the emergence of competitive markets viewing the provision of telecommunications services as a business rather than a public good. Thus, the ITU International Telegraph and Telephone Consultive Committee (CCITT) which handles international accounting settlements, has begun consultations in this regard, though no decisions are expected before 1993 (Staple, 1991).

Public traffic: patterns and factors

It was hinted already in the foregoing discussions that international telephone communications has become a profoundly growing form of communications. In 1990 global traffic reached 30 billion call minutes, and it is growing at an annual rate of 18–20 per cent (Staple, 1991). With the introduction of computer and fax communications international telecommunications have become dominated by telephony at the expense of the more traditional forms of international communications, namely telegraph and telex. Thus, the share of telephony in US international traffic increased from 43.3 per cent in 1975 to 93.5 per cent just 14 years later, in 1989 (Kellerman, 1992a).

The change from telex to telephone as the predominant mode in international telecommunications has been recently studied by Rietveld *et al.* (1992). General declines in telex were evident after 1987, led by rich countries. In most countries the telex system displayed higher growth rates than telephone systems between 1965–80. Exceptions were the US in which declines were registered as of 1982, and Germany and France where telephones grew faster than telex. Generally also the time of peak telex use (1986) leads peak ownership (1987).

International calls may be measured in at least two ways, namely by the number of calls and by the number of call minutes. Annual statistics by telephone companies and PTTs use either one of the two or both. Staple and Mullins (1989a) advocated the use of call minutes as a standard economic measure, since this type of measurement is general to all transmission technologies, and since it is in line with billing procedures. However, detailed comparisons of US outgoing and incoming calls by the two measures showed that both measures have some validity since the transmission of messages differs in its length when using voice, fax or computerized communications (Kellerman, 1992a). Also, minute length reflects supply, whereas the number of calls represents demand, notably in the household sector, where the number of contacts is an important measure, especially since the call length is not planned and is normally determined only after the completion of a call (Kellerman, 1990).

International traffic may be divided into several forms or patterns as far as the geographical destinations of calls to the most frequently called countries are concerned (Kellerman, 1990). One may be termed the global or world economic, presenting strong ties with the leading industrialized nations, the G7 countries. A second pattern would be a traditional or conservative-cultural one, showing preference for contacts with neighbouring countries and countries with language and cultural similarity. Two additional patterns could be a mixture of the two first ones, and a completely open pattern which is not presented as of yet by any country. Another pattern, characterizing many LDCs, accentuates contacts with previous colonial powers.

The dominance of world cores

The traffic between Europe, America and Japan accounts for 75 per cent of global traffic (*The Economist*, 1991). Staple (1991) provided some more details on the traffic volumes about these cores. One quarter of the global traffic is generated by the US. Thus, the most heavily used telecommunications routes world-wide in 1991, with over one billion call minutes in each, were the lines from the US to its most frequently called correspondent countries, namely Canada, Mexico, and the UK. Almost two-thirds of European outgoing international telecommunications is generated by three countries, Germany, the UK and France. Japan, on the other hand, presents much smaller volumes of international telecommunications, probably representing linguistic and daily time differences. In 1990, Japan's 55 million telephone lines generated about the same international traffic as The Netherlands with just seven million lines, or about twice that of the two-and-a-half million lines of Hong Kong (excluding calls to China).

Frequent calling to G7 countries is evidenced by many, though not all, countries presented in Table 7.2. For several countries, such as Germany and Sweden, the priority given to non-G7 countries is explained by the cultural pattern, to be discussed in a following subsection. In other countries, such as Singapore and Hong Kong, a mixed pattern of cultural-linguistic and economic ties is presented (Kellerman, 1990). Using AT&T's *The World's Telephones*, which lists the ten most frequently called countries, it was possible to note that of the 18 countries presented in Table 7.2 only the US has had all other G7 countries at the top of its list, reflecting its global centrality in information distribution. Its list further identifies strong bonds with countries in several continents, resembling to a small degree the open pattern. Most countries seem to have 4–5 G7 countries on their lists. Which G7 countries are included changes with the regional location of each country. Thus, in the Pacific Rim priority in decreasing order is given to the US, Japan and Canada, whereas in Europe the order is Germany, France/UK, Italy, and the US. Generally, the percentage calls to G7 countries, whether close or far, is increasing (Kellerman, 1990).

Table 7.2 The three most frequently called countries for selected nations 1977–87

Country	First	Second	Third
Belgium	France	Netherlands	West Germany
Brazil	US	Italy (1977); Argentina	Argentina (1977); Italy/Germany (West)
Canada	US	UK	Italy (1979–80); Germany (West)
France	Germany (West)	Belgium (1979); UK/Italy	Belgium (1980); Italy/UK
Germany (West)	Austria	UK (1986); Netherlands/Switzerland	Netherlands (1979; 82; 85) Italy/Switzerland
Hong Kong	Taiwan; China (1986–?)	Japan; China (1985) Taiwan (1986–?)	Macau/US; Japan (1985–86)
Italy	Switzerland (1977–80); Germany (West)	Germany (West) (1977–80); Switzerland (1981–83); France	France (1977–83); Switzerland
Japan	Korea (1977–80); US	US (1977–80); Korea	Hong Kong (1982); Taiwan
Korea (South)	Japan	US	Taiwan (1979); Saudi Arabia (1985–86); Hong Kong
Mexico	US	Canada	Guatemala (1980); Colombia (1987); Spain
Netherlands	Germany (West)	Belgium	UK
Singapore	Indonesia; Hong Kong (1979–80)	Indonesia (1979–80); Hong Kong (1979); Japan	Japan (1977–81; 83) US (1983); Hong Kong
Spain	France	Germany (West); UK (1986; 87)	UK; Germany (West) (1986; 87)
Sweden	Denmark/Finland	Denmark/Finland; Norway (1986)	Norway; Finland (1986)
Switzerland	Germany (West)	Italy (1977–82); France	France (1977–82); Italy
Taiwan	Japan	Hong Kong/US	Hong Kong/US
UK	Germany (West) (1977–79); US (1980–83)	France (1977); US (1979); Germany (West) (1980–83)	US (1977); France (1979–83)
US	Canada	Mexico	UK

Data source: *The World's Telephones*, 1977–88.
Source: Kellerman, 1990.

Table 7.3 The ranks and shares of G7 countries in phone calls made from G7 countries in 1986

Origin Country Destination Country	US	Canada	UK	West Germany	France	Italy	Japan
UK	– 	1 90.3	1 24.2	8 5.4	7 6.2	X 	1 25.7
Canada	1 29.4	– 	7 3.8	X 	X 	X 	X
UK	3 9.5	2 3.4	– 	2 11.0	3 10.7	4 8.7	5 7.2
West Germany	4 5.2	3 1.4	2 11.0	– 	1 14.5	1 23.4	7 4.5
France	6 2.5	8 0.3	3 9.3	X[1] 	– 	2 17.3	9 2.0
Italy	7 2.2	10 0.3	5 4.7	6 8.6	2 10.7	– 	X
Japan	5 3.6	X 	X 	X 	X 	X 	–

X = not among the top ten
Data source: *The World's Telephones*, 1977–88.
[1] Ranked fourth in Staple and Mullins, 1989b.

Source: Kellerman, 1990.

Looking at the ties of the G7 countries among themselves (Table 7.3), reveals the US again in a leading position; it is the only country which both has all the other G7 countries on its list, and is on the list of all others. The US was ranked first in three other countries, namely in Canada, the UK and Japan. These countries represent a neighbouring nation (Canada), and the two other leading countries in the global capital economy (the UK and Japan). The US is shown to lead this economy as far as information flows are concerned. Furthermore, Japan and Canada did not appear on any list except for that of the US and in Canada's case also on the list of the UK. By the same token, the US ranked low on continental European lists, whereas France, Italy and Germany ranked high in the US.

Second in importance seems to be Germany, which was rated first by its two European neighbours, France and Italy, and second to fourth by all the others except Japan (it was ranked seventh there). However, Germany's own list was completely different. No G7 country ranked first, two ranked low (Italy and the US), three others were not on the list at all (Japan, Canada and France), leaving only the UK in the second rank (and Germany has the same rank in the UK). Germany may thus serve as the most important correspondent country for many European nations, but its own telecommunications map is more culturally shaped as we will note later. The UK which was ranked second to fifth by most G7 countries and which ranked them accordingly, seems to serve as a bridge between continental Europe and the global 'three-legged stool'. France, Italy, Canada and Japan followed the UK in a descending order of ranking. Interestingly, similar rankings were revealed when corresponding matrices were constructed for trade and tourism among the G7 (Kellerman, 1990).

Longitudinal studies of geographical destinations of international telephone calls were performed for the US (Kellerman, 1992a; 1992b) and Israel (Kellerman and Cohen, 1992). In the US, Canada and Mexico have been traditionally excluded from FCC reports, but the two countries have always topped the list of most frequently called countries from the US, in this order, receiving some 32.3 per cent of the US outgoing calls in 1988. The other five G7 countries have consistently followed the UK

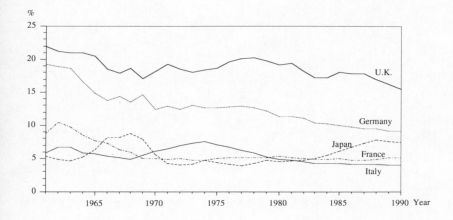

Figure 7.4 The shares of the five leading industrial nations in US outgoing international telephone calls, 1961–90. Data source: US Federal Communications Commission (FCC), 1961–90.

with Germany always ranking first (Figure 7.4). The seniority of the UK attests to the special status of Britain in American information exchanges, both socially and economically. Germany was always second, but its share declined from 19.3 in 1961 to just 9.1 per cent in 1990, or a decline of about 50 per cent. The shares of Japan and Italy rose during the 1960s and 1970s respectively, for political–military reasons. The repeating increase in the share of Japan in the 1980s reflects its growing role in the US economy.

The countries ranked 6–15 on the changing US list of most frequently called countries belong to four world regions (Figure 7.5). Noticeable is the rise in the share of the Pacific Rim during the Vietnam War, in the 1960s, and the more current rise of the region, representing stronger economic and social ties with the US. This rise was coupled with a more modest rise in the share of Latin American countries and a corresponding decline in the share of European nations. The Middle East has been represented by Israel, with Saudi Arabia joining in for several years during the late 1970s, and Iran was on the list during the

174   Telecommunications & geography

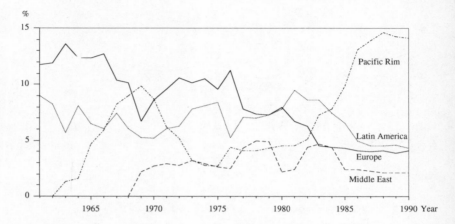

Figure 7.5 The shares of four world regions in US outgoing international telephone calls to countries ranked 6-15, 1961-90. Data source: US Federal Communications Commission (FCC), 1961-90.

early 1980s. The list presents, therefore, a mixture of world-wide economic, social and political interests of the US. Interestingly enough, the list of most frequently called countries for 1990 was arranged exactly in this geographical order of world regional priorities (Table 7.4).

In Israel, the list of most frequently called countries is headed by the US and European countries since 1951, and as of 1973 these two blocks amounted to over 90 per cent of the Israeli international traffic. As of 1983 the share of North America surpassed that of Europe, with the share of the US alone being over 40 per cent. This pattern attests to the intensive ties of Israel with the Western cores rather than with its neighbouring countries (Kellerman, 1991b; Kellerman and Cohen, 1992).

Cultural patterns

Germany and Sweden are prime examples of countries which adhere to the cultural pattern in their communications ties (Table 7.2; Kellerman, 1990). In 1987 Germany had Austria at

Table 7.4 The shares of the 15 most frequently called countries from the US 1990 (in % by number of calls)

Country	Share	Region	Share
UK	15.5		
Germany	9.1		
Japan	7.4	G7	41.1
France	5.1		
Italy	4.0		
S. Korea	3.5		
Taiwan	3.2		
Hong Kong	2.6	Pacific Rim	14.1
Australia	2.4		
Philippines	2.4		
Brazil	2.2	Latin	4.3
Colombia	2.1	America	
Switzerland	2.1	Europe	
Israel	2.1	and	6.2
Netherlands	2.0	Israel	
Total	65.7		

Note: Total excludes Canada, Mexico, Bermuda, and the West Indies.
Data source: US FCC, 1990.

the top of its list, followed in order by Switzerland, Italy, East Germany, and The Netherlands. All these countries are either German-speaking, contain a German-speaking minority or speak a language close to German. The first G7 country on Germany's list of most frequently called countries, the UK, appears sixth. This pattern was consistent for Germany for the whole period 1977–87, and a similar one was presented by Sweden. The three other Scandinavian nations (Denmark, Norway, and Finland) top Sweden's list (in changing orders), followed by Germany in fourth place. While one may argue that for both Germany and Sweden these top-ranking nations are also the most adjacent, in no other case of multicountry adjacency has there been such a dominant priority given to countries of similar language and culture. In both Germany and Sweden the total share of the culturally close countries in outgoing international telephone calls has been declining throughout the second half of the 1980s. Thus, in Germany it declined from 51.4 per cent in 1984 to 46.8 per cent in 1987, and in Sweden from 52.7 per cent in 1983 to 48.7 per cent in 1986.

Colonial imprints

Some imbalance may be identified for the countries most frequently called from nations located outside the three world core regions (Kellerman, 1990). One may repeatedly find core countries as the most frequently called ones, but the calling nations rarely appear on the equivalent lists of the called core countries. This may stem either from differences in population sizes between small periphery nations and large core ones, but more often it relates to the traffic volumes generated by and for developing nations, which are small compared with the contacts made among core countries.

African and Asian nations which were under colonial rule until some time in the twentieth century maintain the former colonial nation as the most frequently called nation. This reflects both linguistic-cultural and economic ties. It was found that the level of interaction with a former colonial power is 3.25 times higher than with other countries (Rietveld and Rossera, 1992). Other most frequently called nations are neighbouring countries and other core countries. Latin American countries which received their independence from Spain and Portugal before the twentieth century, but still speak the previous colonial language, present a different pattern. It is the US, rather than Spain or Portugal, that has been ranked first, and in very high percentages, in almost all Latin American countries. This reflects the status of the US as the strongest economic power in the western hemisphere and its being the haven for many emigrants from Latin American countries. It is interesting to note that the US ranked first even in Cuba, with 72.9 per cent of calls in 1987 going to the US and just 3.0 per cent going to the Soviet Union. Spain has ranked variably in Latin American nations as the third to seventh most frequently called nation.

Two other interesting regions are the Middle East and Eastern Europe; the former because of the Arab culture and language shared by most nations in the region and the latter because of the Soviet political predominance which prevailed until recently. In Middle Eastern countries one would expect Egypt to be ranked first, being the largest and culturally leading Arab nation. However, it is in fact Saudi Arabia that has taken the first rank in many Arab countries. This preference might be related to the large number of non-Saudi Arabs working in Saudi Arabia. Egypt, for its part, has the US topping its list as the most frequently called country, and Western countries are ranked high by other Arab countries, as well. In the Gulf states, countries such as Pakistan and the Philippines appear too, since citizens of these nations have been employed in large numbers in the Gulf nations.

One might have expected East European nations to have the Soviet Union ranked first in the 1970s and 1980s. However, this has been the case for Bulgaria only. Interestingly, Bulgaria was ranked first in the Soviet Union in 1986–87. This mutual trend might represent both the existence of especially strong political and economic bonds between these two countries, and/or close similarities in culture and language. Other Eastern European nations had European nations topping their lists with West Germany leading, and the Soviet Union appearing only lower in their lists. The Soviet list, however, ranked Eastern European nations first.

International telecommunications and other international movements

The dependence of the international movement of information via telecommunications in any country on the movement of people, commodities, and capital may be tested for three aspects: (1) annual growth rates in traffic at large; (2) geographical variations in traffic volumes by destination countries in any given year; and (3) variations along time in traffic to specific countries. Detailed longitudinal studies using this international movement model have been pursued for the US

(Kellerman, 1992a; 1992b), and for Israel (Kellerman and Cohen, 1992). Partial examinations were conducted for several other countries (Kellerman, 1990), for The Netherlands (Rietveld and Janssen, 1990), and were proposed for Portugal (Gaspar and Jensen-Butler, 1990).

The average annual growth rate for American international telecommunications 1962–89 was 25.6 per cent. Regressing the annual growth rates 1962–88 against the growth rates of the two-way movements of capital, people, and commodities yielded low results (R-square = .32). The best results were received for the period 1965–79 (R-square = .76). This finding was interpreted as reflecting a technological factor. The period before 1965 (when satellites were introduced) was typified by low supply, whereas the period 1965–79 was typified by growth in both supply and demand, through the introduction of new transmission and dialling technologies yielding lower prices and higher service qualities. In the 1980s supply exceeded demand again, especially as of the mid-1980s, when it was estimated that each of the US major carriers had the capacity to transmit the whole US demand for international services (Staple and Mullins, 1989a).

The most important variable in the explanation of growth was foreign visitors to the US, representing a blend of social and economic ties with the US. For Israel, annual growth rates for the period 1952–88 were analysed, with the two-way movements of people and commodities serving as independent variables. The best explanation was achieved for the period 1975–87 when exports and imports were lagged two years after telecommunications (R-square = .88). Here too the period of most intensive technological improvement in telecommunications infrastructure was best explained by the model.

Variations in traffic volumes to most frequently called countries in each year separately were also well explained by the international movement model. In the US the levels of explanations were higher than R-square = .96 for every year 1961–89. The most important variable was outgoing tourism from the US, and the general order of explanation by variables in descending order was people, capital, and commodities. This attests once again to the impact of households, or social calls, which stem from visits to foreign countries, blended with the impact of business visits.

The increasing importance of global capital markets is expressed in the more important role of capital exchanges compared to that of commodities (Kellerman, 1992a; 1992b). Similar analyses for Israel yielding high results were obtained as of 1957, though here exports were generally found to be the crucial explanatory variable (Kellerman and Cohen, 1992). In The Netherlands a similarly high explanation by exports was found for 1983 (Rietveld and Janssen, 1990). In partial analyses for 1985 and 1986 for France, West Germany, Italy, Netherlands, Switzerland, and the UK high results were obtained for all countries except for Germany, given the unique nature of the geographical distribution of its international telephone calls. In all countries, except for the UK, it was exports or imports which served as the most important variable, rather than tourism (Kellerman, 1990). The wide use of the telephone in the US and its being an immigration country may explain the predominance of social variables there. British data, on the other hand, require further analysis before decisive conclusions can be made.

The use of the international movement model for explanations of trends in the changing number of calls to specific countries has proven more doubtful. Results were mixed in both the US and Israel, namely the obtained levels of explanation were different from country to country, as well as the leading variable, reflecting the changing nature of relations with foreign countries. The mean R-square coefficient for analyses of 11 countries most frequently called from the US over the years was .74. The coefficients were high for France and Italy where the leading variable was outgoing tourism to these two popular countries. They were also high for Switzerland with American financial investments as the leading variable, or for Israel with exports leading. On the other hand, the explanation was low for the UK with its complex ties with the US, as well as for Venezuela (Kellerman, 1992b).

For Israel similar trends could be observed. Thus, the leading variable for calls to Germany for the period 1951–63 was tourism from there, and low explanatory results were obtained for the UK, France, The Netherlands, and Italy. For the period 1973–88 it was exports and imports which provided the best explanation for the annual variation of calls to Switzerland, while the size of the Jewish community led the explanations for the US and France.

Dedicated international networks

Dedicated international networks, whether leased or private, are no less complex than public ones, but information on them is scant, since they are private, and given their nature of continuous operation, measuring the number of calls and call minutes may turn out difficult or impossible. Network structures are not different than those used by public networks (see Figure 2.2; see also Langdale, 1989a, p.508; and Bakis, 1987). The uses that can be made of dedicated networks are highly varied, and attention here will be given to the three major ones, namely the networks developed by and for transnational corporations and banks; offshore data processing; and electronic mail networks. Before moving to these uses, some more general aspects of dedicated networks have to be addressed.

Transborder data flows (TBDF)

Dedicated networks are frequently labelled under the term transborder data flows. The term was originally coined by the Organization for Economic Cooperation and Development (OECD) in 1974 (Jussawalla and Cheah, 1987, p.27). It relates to 'the flow of digital information across borders for storage or processing in foreign computers and reflects the interests of governments in regulating or otherwise intervening in the free flow of such exchanges for a variety of reasons including economic, social, political, and cultural' (Smith and Healy, 1987, p.67). Thus, by the mid-1980s at least 22 nations restricted TBDF in one way or another (Smith and Healy, 1987).

From an economic perspective, European PTTs felt threatened by the possible competitive penetration of private enterprise (Jussawalla and Cheah, 1987, p.12). Potentially, contemporary technologies permit the re-sale of leased lines for voice telephony, so that dozens of conversations may simultaneously take place on one line. As far as national economies are concerned, many nations, both developed and developing ones, apprehended growth in TBDF. It may discourage the development of domestic data processing, a consideration which led Brazil to firmly restrict foreign data processing, even for foreign companies operating in the country, such as the Swedish Volvo company (Langdale, 1989a).

Politically, there are countries which view TBDF as a loss of national power and sovereignty, notably to US firms. This may be the case even in commercial enterprises, such as having a European airline reservation system run from a computer system located in the US. Such concerns have been strongly raised in Canada (Smith and Healy, 1987). TBDF also involves a social problem, namely that of privacy violation. Sweden has become known for its strict policy in this regard, requiring an export license for any data which may identify individuals (Keen, 1988, p.171).

Transnational corporations (TNCs) and banks

It was estimated in the mid-1980s that 90 per cent of the volume of TBDF was intracorporate (Jussawalla and Cheah, 1987, p.4). Leased lines by transnational corporations facilitate international production, processing, and management, and were analysed as such by Langdale (1989a). Major industrialized countries compete for the location of hubs of leased networks, since they may attract additional related economic activity. Britain, Ireland, Hong Kong, Singapore, Japan, and Australia are prime examples, for both a competitive and less regulated supply of leased networks, as well as high levels of demand for such networks by TNCs, mainly American ones, which are looking for convenient locations for telecommunications headquarters serving their global business activity.

Leased networks are useful for large companies, notably in high-tech industries and financial services. They permit TNCs to internalize a substantial share of their international telecommunications traffic. As the lines are leased for a flat fee, this is of special importance, since the cost per unit use decreases as usage increases. Dedicated networks are, thus, considered part of the competitive advantage of companies, the lines are felt to be more secure, and they may eventually permit leasing companies to enter the market of public networks (Langdale, 1989a). Leased lines are also vital for inter-company communications, whether industrial ones or interbank networks, such as SWIFT (Langdale, 1989a).

Offshore back offices

Offshore back offices refer to office activity, normally data processing, performed in foreign countries, so that documents are shipped by air from one country to another for processing and the processed information is then transmitted back via telecommunications. This kind of activity is similar to the relocation of data processing from large cities to rural America discussed in Chapter 5. In fact offshore data processing may compete with similar facilities located in rural America, but this is not always the case. Whereas labour wages in the Caribbean and Asia, the two preferred regions for offshore activity, range from one-fourth to one-fifth of US wages, the turnaround time from rural America is 12 hours compared to Barbados where it takes 48 hours. American workers are also preferred for government jobs (Howland, 1992).

Offshore data processing may also be interpreted as a phase in the development of service functions similarly to the last phase in the *product cycle* typifying the electronics industry (see e.g. Scott, 1982). Thus, when data processing becomes routine and does not require high skills, and when transportation and transmission facilities become available it will be moved to foreign countries offering lower wages. Additional requirements would be low taxes (Wilson, 1992), and the availability of a literate workforce, mainly women, and preferably English speaking ones (Wilson, 1992; Howland, 1992).

It was estimated in 1989 that the world-wide offshore industry was rather small, employing some 10,000 people (Castells, 1989, p.164), with 150 back offices located in the Caribbean and a dozen in Ireland (Wilson, 1992). Offshore back offices may be found in the Caribbean (Jamaica, Barbados, and the Dominican Republic), Asia (India, Philippines, South Korea, and China), and Europe (Ireland, Scotland) (Howland, 1992; Wilson, 1992). Whereas American firms were the first to establish contract or in-house offshore back offices, they were recently joined by Canadian, British, and Japanese ones. These firms normally deal with insurance, aviation, publishing, market research, and financial services (Wilson, 1992).

It is generally agreed that the industry is vulnerable to the current technological improvements in image processing, scanning and bar coding which may either eliminate, make more efficient, or require technological skills for data processing. Such changes may return the remaining activity to company headquarters in the industrialized countries (Castells, 1989; Wilson; 1992; Howland, 1992).

Electronic mail networks

Electronic mail networks are public networks which permit the transmission of files and mail. Such networks are mostly interconnected national networks, which might be operated by PTTs, telephone companies, commercial companies or scientific organizations. 'Networks may be not only communities of convenience, but also communities of interest' (Quarterman, 1990, p.21). Thus, they may be used by a variety of organizations and persons, ranging from academicians to Amnesty International. Obviously, TNCs may operate their own private e-mail networks. We noted in Chapter 2 the problems involved in the adoption of e-mail, but it was claimed that the use of e-mail may increase efficiency (Sproull and Kiesler, 1991, p.23). The introduction of easy-to-use fax reduced the attractiveness of e-mail even further. Even the most noticeable advantage of electronic mail over mail, the speed of message production, has been eliminated with the integration of fax and computer technologies, so that messages arriving by fax may be displayed and stored by computers.

The world's largest e-mail network is *Internet*, linking several million people through 750,000 computers ('hosts') (*The

Economist, 1992). In 1989, the British Telecom Gold e-mail system was estimated to have 140,000 users, compared to 350,000 fax machines in Britain at the same time. It was further estimated for the end of 1988 that there were some 11 million electronic mailboxes world-wide as against 7.8 million fax machines (Hall, 1991). Given the tremendous diffusion rates of fax machines one may safely assume that fax has become the preferred technology for electronic message transmission.

National electronic networks are usually organized around 'backbones', which are electronic super-highways, or high-capacity transmission links, running along major routes, and into which lower capacity and local lines are connected, in hierarchical or mesh forms (Batty, 1991). An alternative organizational form could be a tree-like one (Figure 2.2), which is the case for the BITNET system, where each joining host has to agree to the connection of a future joining host to the system through it (Kellerman, 1986a; 1986b). National systems are interconnected by gateways, sometimes referred to as internets (Batty, 1991), such as Rome and New York in the BITNET system. Major gateways for inter-network connections are located in Cambridge, MA, and in various American universities, as well as in Amsterdam, Melbourne, Seoul, Tokyo, and Geneva (Quarterman, 1990).

The most widely known e-mail system in the academic world is BITNET (Because It's Time Network) established in 1981 when Yale University and City University of New York were linked. The system spread very fast and it has become a major tool for communications among academicians world-wide (Kellerman, 1986a; 1986b). In 1990 the system linked 2,300 hosts at several hundred sites in 32 countries, and it consisted of four networks: BITNET in the US and Mexico; NETNORTH in Canada; ASIANET in Japan, Korea and the Far East; and EARN in Europe. The gateway for the European system is located in Montpellier, France (Figure 7.6; Quarterman, 1990, p.430). The system is also connected to other national e-mail systems such as the British JANET (Batty, 1991).

Figure 7.6 The EARN network (25 April 1988). Source: Quarterman, 1990, p.430.

Conclusion

International telecommunications might well be considered the geographical dimension which has been transformed most during the three decades since the introduction of satellites. A variety of instantaneous services and channels have become available for both businesses and households. The exposure of households to direct and instant international telecommunications may well be the most significant change in telecommunications services available to households so far.

International telecommunications is also the front in which

major future changes should be expected from organizational and economic respects. If current trends continue, then prices may fall with changing settlement agreements, and a possible broad introduction of international competition, similar to the one prevailing in the airline industry, may be revolutionary in this regard.

In a world with a growing global and internationally integrated economy, telecommunications means become crucial highways for information of all kinds. It remains to be seen to what extent existing national structures of telephone services will react to the technological, economic, and social challenges posed by contemporary and future telecommunications means.

Chapter 8
Conclusion

This chapter will attempt to provide a concluding geographical perspective on telecommunications, moving from the past through the present to the future. There are two dimensions which have been repeatedly raised in the preceding discussions, and which lead from past developments to present processes. First are geographical concentration and dispersion patterns and processes related to telecommunications at various geographical scales, and second is the leading role played by the US in many geographical and other aspects of telecommunications.

Present processes and patterns may lead to observations and speculations on future developments in two ways. It is possible to use the current geographical portrait of telecommunications as a building block for a possible future emergence of a theoretical framework for the geography of telecommunications. Furthermore, it is intriguing to look at contemporary developments in the telecommunications industry and, using direct or analogous observations, to speculate on future geographical patterns which maybe nesting within current ones, as well as new patterns which may emerge in the future.

Telecommunications, concentration and dispersion

We have noted the concentration tendency of telecommunications facilities as well as of telecommunications-related economic activities at various scales. These were teleports and CBDs at the local–urban scale; the global urban hierarchy at the urban–global scale; interregional inequality at the regional scale; and national and international concentrations of services and traffic.

On the other hand, we have observed various dispersion patterns again at various scales. This has been most noticeable in the service industry, where back offices may transfer to suburban areas at the local scale, to peripheral regions at the national scale, and to offshore locations at the international scale.

Comparing the extent of these concentration and dispersion movements it seems clear that telecommunications facilitates foremost concentration. Concentration patterns and processes have been shown to be strong and of an accumulative nature, whereas dispersion patterns are weak and vulnerable to technological and other changes. Furthermore, the decentralization processes of economic activities actually amount to an outgrowth of the centralization trends. In other words, *if there is no concentration first there will be no following dispersion*. It is obvious that decentralization may lead to an accumulation of controlling power in major cities (Sassen, 1991). However, the 'driving force' for decentralization is earlier concentrations of power and controlling channels, such as telecommunications. Decentralization may, therefore, often constitute a spillover effect of the centralization processes, which may yield additional concentrations of power and controlling means.

This assertion applies directly and indirectly. Directly, the concentration of heavy telecommunications-dependent activities in large cities may bring about increasing real estate prices, possibly leading to a dispersion of office activities to suburban and peripheral locations. Indirectly, a growing demand for low-wage telecommunications-related employment makes peripheral areas more attractive, given the lower wages they may offer. In many cases the direct and indirect forces may operate jointly. Similar processes may apply to the dispersion of manufacturing. If a city cannot offer controlling means *vis-à-vis* telecommunications means, well-developed producer services, and centralized headquarters, the lower wages and real estate prices in peripheral regions will not suffice for a decentralization of production.

The more flexible migration patterns which telecommunications services may encourage are also related to these effects. In other words, major concentrated employment foci may attract workers from distant locations who will be able to be in touch with their more dispersed families by using telecommunications means. The concentric pattern of telecommunications is not merely a geographical one, but a sectoral one as well. It was reported that 90 per cent of AT&T's business toll revenues was generated by merely 12 per cent of its business customers, and it was for 300 corporations to generate 25 per cent of the company's total business revenue (Morgan and Pitt, 1989). By the same token, 60 per cent of data communications traffic in the UK is generated by 300 large companies, in the US 50 per cent of carriers revenues are produced by 4 per cent of the users, and in Norway 40–50 per cent of data traffic is generated by 25 companies (Hepworth, 1990, p.193).

Telecommunications-related concentration trends of economic activities include capital markets, information activities and producer services. These activities usually emerge where capital has accumulated through previous developmental phases, namely industrial production. Telecommunications may facilitate the transition from a manufacturing accent to a service one. It permits a centralization of capital markets and decision-making, while industrial production may disperse to peripheral regions or NICs, and connected to companies' headquarters via telecommunications. Telecommunications and air transportation may thus be viewed as the prime channels of the post-industrial economy, equivalently to the role of railways and maritime transportation for the industrial economy. Telecommunications has not emerged as a substitute for the movement of people and commodities, but rather as a complementary, dependent and sometimes even catalyzing transmission mode.

The considerable concentration effects coupled with the more restricted dispersion processes imply the existence of telecommunications-related cores and peripheries, which are organized along hierarchies at different scales. The larger urban concentrations in the global urban hierarchy constitute cores, and cores at a larger scale are the world core regions of North America, Western Europe, and the Pacific Rim. Moving along the urban hierarchy, cores are first in adopting new technologies and organizational changes, and they are also first to reap the fruits of these innovations. The spatial organization of society reflecting an integration of advanced telecommunications technologies may not be radically different than that of the industrial age, but the *geographical scale* of economic and social activities and ties maybe much wider, at the domestic as well as at the global levels. World cities, for example, have become areas of increased economic activity requiring larger forelands and sometimes also larger hinterlands.

Telecommunications has been variously described as spatially neutral, but this is so only potentially. As developments for several sectors, scales, and countries attest, telecommunications has been first a technology contributing to concentration processes, followed by some decentralization. For the metropolitan level, Castells (1989, p.167) stated that 'it is the dialectics between these processes of centralization and of decentralization that fundamentally characterizes the new spatial logic'. If the widening and the intensification of external information ties, coupled with concentration and decentralization trends amount to a new spatial logic, then it applies to all scales, namely the regional, national and global ones.

American leadership in telecommunications

The US has been striking in the preceding chapters as a leading core and concentration for almost every aspect in telecommunications. These aspects may be classified into three general categories, namely the innovation, use, and organization of telecommunications systems. It has been in the US where most technological innovations in telecommunications took place since and including the inception of the telegraph in 1837, and that of the telephone in 1876. It has further been for the US to become the first information and service economy (Kellerman, 1985), in the emergence and prosperity of which telecommunications technologies play a crucial role. The American society has also been fast and extensive in the diffusion and adoption processes for the various telecommunications technologies. This has been shown once and again for transmission media (e.g. the telephone, the television), for networks (e.g. satellites, fibre-optics), for systems (e.g. digital equipment), and nodes (e.g. teleports).

The massive adoption of telecommunications technologies and

devices has been coupled with most intensive usages made of them. Thus, American society leads in the number of telephone calls per subscriber, as well as in call-lengths. The US has further been shown to be leading in international traffic to and from the US, in both public and dedicated networks. The increasing importance of households has also typified American international telecommunications. Private ownership of telecommunications has existed until recently almost exclusively in the US, so that it was in the US that telecommunications services have been historically viewed as a mix of utility and business. With the pioneering introduction of competition in telecommunications services the US has turned into a leader in the global reorganization process of the telecommunications industry.

The historical and on-going leadership of the US in telecommunications may be attributed to several factors. Economically, the private ownership of the system, even if monopolistic until lately, coupled with the capitalist structure of the American economy, have provided incentives for continued R&D efforts as well as extensive uses of the telecommunications system. Geographically, the large territorial size of the US, and the development of the US into a nationally integrated economy, have been both assisted by and challenging to the telecommunications industry. Socially, American society has been characterized as a geographically dynamic one, namely the propensity to move to new locations has been higher than in other countries. This geographical restlessness, coupled with the American openness for innovations and new products, have been aided by telecommunications means, and at the same time provided an impetus for the continued development of the telecommunications industry.

The tremendous importance of telecommunications for American life, and the country's leadership in this area have not turned telecommunications in general and the telephone in particular into popular research subjects. It seems that telecommunications have been taken for granted, and the small sizes of telecommunications devices as spatial artefacts have not drawn too much attention to them.

The US has turned into the largest producer and distributor of information in the world, both within the country and globally.

If information is power, then the current shifting seniority in industrial production from the US to Japan has to be viewed properly. The US has enhanced its senior position in information production and transmission, and this was coupled with a geographical shift in industrial seniority moving from the US to Japan. A question that deserves special attention in this regard is whether a global information seniority may economically compensate for a loss in industrial supremacy. We have noted the pressures for a reform in the current old-fashioned international settlement system for telecommunications accounting, which has brought about tremendous deficits in the US national account. Beyond this problem there is a broader question whether information is exchanged internationally at too low prices or not (e.g. in the fields of television, capital exchanges, database information etc.). This question may be related to another one, namely whether the competitive structure which typifies the current production and marketing of commodities has already been fully developed in the area of information production, distribution, processing, and exchange.

Directions for a geographical theory of telecommunications

The two parts of this book may offer some initial theoretical observations for the geography of telecommunications. Part 1 was summarized by the telecommunications cycle (Figure 3.9), which focused on the geographical components of telecommunications systems (media, nodes, networks, and flows). It further focused on the cumulative impacts of the uses of telecommunications systems at the micro level, as well as on the resulting demands for additional access and uses of telecommunications services. Part 2 dealt mainly with the interaction between telecommunications systems, on the one hand, and geographical systems at various scales, on the other. The possible impacts of telecommunications on these systems are summarized in Table 8.1. This table may, therefore, be considered as a continuation of Figure 3.9.

At the urban level, telecommunications may facilitate both

Table 8.1 Major geographical impacts of telecommunications

Scale	Leading dimension	Major impact	Leading sector
Urban	Concentration and dispersion	Spatial	Services
Regional	Location	Economic	Manufacturing
National	Concentration	Economic	Services
	Diffusion	Economic	Manufacturing and services
International	Movement	Social?	Households?
Global	Hierarchy	Economic	Services
Universal	Accessibility	Economic and social	Services and households

concentration (in CBDs) and dispersion (into suburban back offices). Its impact is, thus, mainly felt in the service sector, and is expressed spatially in the form of reinforcing or transformation of land uses and the spatial organization of cities. At the regional level, a transition of some permanence is the possible location (or relocation) of manufacturing plants in peripheral regions, which may strengthen the economic base of such regions (the relocation of services was shown to be more vulnerable to technological change). At the national level, two processes were mentioned. One is the possible heavy concentration of services in major urban areas resulting from privatization and deregulation in national ownership patterns of telecommunications services. The other related to the diffusion process of innovations, which for telecommunications was shown to operate on a hierarchical basis, so that manufacturing and services located in larger urban areas enjoy technological improvements earlier than those located in smaller towns.

At the international level, or for the interrelationships among countries, the movement of information was related to other international movements. The role of households was shown to be of importance for the US and possibly also for the UK. Globally, or for the world seen as one unit, the emergence of a global urban hierarchy was noted, specializing in capital markets and information services. Last but not least, there is also the universal level, which relates to a more general geographical aspect, namely accessibility. This latter aspect may be radically changed through the current rapid diffusion of placeless cellular telephony into both businesses and households.

Castells (1989, p.348) proposed a major geographical trend for the information society: 'the historical emergence of the space of flows, superseding the meaning of the space of places'. This transition from a space of places into a space of flows implies a restructuring of territorially-based institutions and an emerging meaninglessness of places. It may well be that economic institutions changed, and in at least two senses. They have become global in their *scale of operation* (e.g., the emergence of TNCs), and *sectoral activity* shifted from an almost exclusive exchange of commodities to exchanges of commodities, capital, and information. However, these changes have not become placeless at the institutional level; rather their *geographical loci* has usually remained very traditional and based on the previous phases of economic development, namely they remained located in the world cores and in major cities in them. Truly, telecommunications has turned New York, London, and Tokyo into more prominent centres than they were before, and at the expense of other leading cities (such as Paris). However, this very process has made places even more important than in the past as centres for economic activity and decision-making. Even the emergence of the new world core in the Pacific Rim and its leading hubs may be interpreted as a result of capital accumulation which began in the industrial era.

There is, therefore, a difference between what telecommunications may *potentially* facilitate, on the one hand, and *actual patterns* shaped by telecommunications jointly with a variety of forces, on the other. Theoretically, telecommunications may permit limitless flows of information globally, without consideration of natural as well as human-made barriers. Realistically, however, not only are there barriers which shape imbalanced flows of information, but demands and capital for the creation of supply are not homogeneously distributed, and were brought about by previous phases of development, at the various geographical levels.

At the individual level, telecommunications may involve delocalization effects, so that people become less attached to places of residence. However, these too are not placeless processes, namely attractivity varies from place to place and population is more attracted, as in the past, to places which have more to offer economically and socially. Another delocalization process involves the global information provided by television, notably cable programming, but this is beyond the scope of our discussion here.

Telecommunications and geography: future scenarios

The objective of this book was *not* to delve into futuristic forecasting. However, it still seems in place to conclude this chapter with some comments on possible future scenarios. A process which is already in effect is the miniaturization of cellular telephones coupled with falling prices. As was hinted in the previous section, this may turn the term 'accessibility' obsolete. However, the major impact of a possible universal availability via cellular telephones may be temporal rather than spatial, namely that the pace of business and social relations will become faster, and that the separation between leisure time and business time will become more difficult. Geographical flexibility for individuals may increase, but it is difficult to foresee any major changes in the spatial organization of society.

Another trend which is only in its early beginnings is the unification and centralization of information services. Currently, information services are split between radio stations, television broadcasting companies and services, cable television, one or more telephone companies, and separate or private computer services. ISDN technology attempts to provide broadband services for telephony, cable television, and data communications for corporations. The spread of ISDN is constrained, to a considerable extent, by the availability of fibre-optics networks, which do not yet reach most households. However, one could look at the contemporary telecommunications industry as being in the middle of its developmental path, in terms of the organization and distribution of services, when computers are compared to electric dynamos, another 'general purpose engine' (David, 1989). The development of electricity began with independent dynamos being replaced, later on, by central electric power stations, functioning as a utility. The American RBOCs were reported to enter the British cable television business, and granted permission to offer cable television services over the telephone lines, in a 'broadband network' (*The Economist*, 1991). Combined telephone and television services would make feasible the heavy investments required to rewire American households with fibre-optics.

If one can really draw a lesson from electricity and from current technologies and business desires, it might well be that the currently separate information and telecommunications industries will be unified, and customers will receive all information services, including computer services, as a utility, thus eliminating all the currently used independent computers of various sizes. Since the US has been shown to lead the innovation of technology and organization in telecommunications, the possible widening of services provided by local telephone companies may signal developments in this direction. The potential development of all-purpose central information-telecommunications systems do not necessarily imply major changes in the spatial organization of society. It does imply, however, a tremendous power concentration by the service providers. It may further imply an increased complexity of international relations, mainly at the cultural and political levels, when information of all types will become available everywhere.

This concluding future scenario should be accompanied by observations made by two scholars in the field of telecommunications and geography. Abler (1991, p.46) commented:

> We will think and teach more effectively if telecommunications and society are viewed as a coevolving complex of processes: telecommunications, societies, economies, governments, human aspirations and human values shape each other, in much the way complexes of living organisms evolved in interaction with each other, each being cause as well as effect of the other's change.

And Hepworth (1990, p.217) stated: 'in my view, the future value of geographical research to public policy debates hinges on whether we can *rapidly* develop a sound understanding of the spatial aspects of the information economy'. It is hoped that this book made a modest contribution in these two directions.

References

Abbatiello, J. and Sarch, R. (eds.) (1987), *Telecommunications and Data Communications Factbook*, New York: Data Communications and CCMI/McGraw-Hill.

Abler, R.F. (1968), 'The geography of intercommunications system: the postal and telephone systems in the United States', PhD dissertation, University of Minnesota, Department of Geography.

Abler, R.F. (1970), 'What makes cities important', *Bell Telephone Magazine*, 49 (2), pp.10–15.

Abler, R.F. (1974), 'The geography of communications', in Hurst, M.E. (ed.), *Transportation Geography: Comments and Readings*, New York: McGraw-Hill, pp.327–45.

Abler, R.F. (1975), 'Effects of space-adjusting technologies on the human geography of the future', in Abler, R.F., Janelle, D.G., Philbrick, A. and Sommer, J. (eds.), *Human Geography in a Shrinking World*, North Scituate, MA: Duxbury Press, pp.35–56.

Abler, R.F. (1977), 'The telephone and the evolution of the American metropolitan system', in Pool, I.d.S. (ed.), *The Social Impact of the Telephone*, Cambridge, MA: MIT Press, pp.318–41.

Abler, R.F. (1991), 'Hardware, software, and brainware: mapping and understanding telecommunications technologies', in Brunn, S.D. and Leinbach, T.R. (eds.), *Collapsing Space and Time: Geographic Aspects of Communication and Information*, London: Harper Collins Academic, pp.31–48.

Abler,R.F. and Adams, J.S. (1977), 'The industrial and occupational structure of the American labor force', *Papers in Geography*, 15, Pennsylvania State University, Department of Geography.

Abler, R.F., Adams, J.S. and Gould, P. (1971), *Spatial Organization*, Englewood Cliffs, NJ: Prentice-Hall.

Abler, R.F. and Falk, T. (1981), 'Public information services and the

changing role of distance in human affairs', *Economic Geography*, 57, pp.10–22.
Ackerman, E.A. (1958), *Geography as a Fundamental Research Discipline*, Chicago: University of Chicago, Department of Geography Research Paper 53.
Adrian-Bueckling, M. (1982), 'Commercial Space Satellites and the problems associated with them', *Universitas*, 24, pp.239–44.
Akhavan-Majid, R. (1990), 'Telecommunications policymaking in Japan: the 1980s and beyond', *Telecommunications Policy*, 14, pp.159–68.
Akwule, R.U. (1991), 'Telecommunications in Nigeria', *Telecommunications Policy*, 15, pp.241–47.
Antonelli, C. (1979), 'Innovation as a factor shaping industrial structures: the case of small firms', *Social Science Information*, 18, pp.877–94.
Aronson, S. (1971), 'The sociology of the telephone', *International Journal of Comparative Sociology*, 12, pp.153–67.
Bakis, H. (1981), 'Elements for a geography of telecommunication', *Geographical Research Forum*, 4, pp.31–45.
Bakis, H. (1984), *Geographie des Telecommunicationes*, Paris: Presses Universitaires de France.
Bakis, H. (1987), 'Telecommunications and the global firm', in Hamilton, F.E.I. (ed.), *Industrial Change in Advanced Economies*, London: Croom Helm, pp.130–60.
Bakis, H. (1988), 'Technopolises, teleports, telecomplexes, telebases . . . telecommunications and sites to be equipped'. Paper presented at the 26th International Geographical Congress, International Geographical Union, Sydney, Australia.
Ball, D. W. (1968), 'Toward a sociology of telephones and telephoners', in Truzzi, M. (ed.), *Sociology and Everyday Life*, Englewood Cliffs, NJ: Prentice-Hall, pp.59–75.
Batty, M. (1988), 'Home computers and regional development: an exploratory analysis of the spatial market for home computers in Britain', in Giaoutzi, M. and Nijkamp, P. (eds.), *Informatics and Regional Development*, Aldershot: Avebury (Gower), pp.147–65.
Batty, M. (1991), 'Urban information networks: the evolution and planning of computer-communications infrastructure', in Brotchie, J., Batty, M., Hall, P., and Newton, P. (eds.), *Cities of the 21st Century: New Technologies and Spatial Systems*, New York: Halsted Press, pp.139–57.
Beesley, M.E. (1992), *Privatization, Regulation and Deregulation*, London: Routledge.

Bell, D. (1973), *The Coming of Post-Industrial Society*, New York: Basic Books.
Bell, D. (1980), 'Introduction', in Nora, S. and Minc, A. *The, Computerization of Society*, Cambridge, MA: MIT Press.
Beniger, J.R. (1986), *The Control Revolution: Technological and Economic Origins of the Information Society*, Cambridge, MA: Harvard University Press.
Berry, B.J.L. (1972), 'Hierarchical diffusion: the basis of developmental filtering and spread in a system of growth centres', in English, P.W. Mayfield, R.C. (eds.), *Man, Space and Environment*, New York: Oxford University Press, pp.340–59.
Beyers, W.B. (1989), 'Speed, information exchange, and spatial structure', in Ernste, H. and Jaeger, C. (eds.), *Information Society and Spatial Structure*, London: Belhaven Press, pp.3–18.
Bezek (Israel Telecommunications Co.) (1991), *1990 Statistical Yearbook*, Jerusalem: Bezek.
Bikson, T.K. and Schieber, L. (1990), *Relationships between Electronic Information Media and Records Management Practices: Results of a Survey of United Nations Organizations*, Santa Monica, CA: The Rand Corporation.
Brooker-Gross, S.R. (1980), 'Usages of communication technology and urban growth', in Brunn, S.D. and Wheeler, J.O. (eds.), *The American Metropolitan System: Present and Future*, New York: Wiley, pp.145–59.
Brooker-Gross, S.R. (1981), 'News wire services in the nineteenth-century United States', *Journal of Historical Geography*, 7, pp.167–79.
Brunn, S.D. and Leinbach, T.R. (eds.) (1991), *Collapsing Space and Time: Geographic Aspects of Communication and Information*, London: Harper Collins Academic.
Buyer, M. (1983), 'Telecommunications and international banking: the political and economic issues', *Telecommunications*, 32, pp.44–52.
Carey, J. and Moss, M.L. (1985), 'The diffusion of new telecommunication technologies', *Telecommunications Policy*, 9, pp.145–58.
Case, D.O. and Ferreira, J.H. (1990), 'Portuguese telecommunications and information technologies: development and prospects', *Telecommunications Policy*, 14, pp.290–302.
Castells, M. (1985), 'High technology, economic restructuring and the urban–regional process in the United States', in Castells, M. (ed.), *High Technology, Space and Society*, Beverly Hills, CA: Sage, pp.11–20.
Castells, M. (1989), *The Informational City: Information Technology,*

Economic Restructuring, and the Urban-Regional Process, Oxford: Basil Blackwell.

Cheong, K. and Mullins, M. (1991), 'International telephone service imbalances: accounting rates and regulatory policy', *Telecommunications Policy*, 15, pp.107–18.

Cherry, C. (1970), 'Electronic communication: a force for dispersal', *Official Architecture and Planning*, 33, pp.733–76.

Christaller, W. (1933), *Die Zentralen Orte in Sueddeutschlan*. Jena: G. Fischer. Baskin, C.W. (tran.) (1966), *The Central Places in Southern Germany*, Englewood Cliffs, NJ: Prentice-Hall.

Christie, B. and Elton, M. (1979), 'Research on the differences between telecommunications and face to face communication in business and government', in *Impacts of Telecommunications on Planning and Transport*, London: Departments of the Environment and Transport Research Report 24.

Clapp, J.M. and Richardson, H.W. (1984), 'Technological change in information processing industries and regional income differentials in developing countries', *International Regional Science Review*, 9, pp.241–56.

Codding, G.A. Jr. (1991), 'Evolution of the ITU', *Telecommunications Policy*, 15, pp.271–85.

Cohen, R.B. (1981), 'The new international division of labor, multinational corporations and urban hierarchy', in Dear, M. and Scott, A.J. (eds.), *Urbanization and Urban Planning in Capitalist Society*, New York: Methuen, pp.287–315.

Corey, K.E. (1982), 'Transactional forces and the metropolis', *Ekistics*, 297, pp.416–23.

Corey, K.E. (1991a), 'The role of information technology in the planning and development of Singapore', in Brunn, S.D. and Leinbach, T.R. (eds.), *Collapsing Space and Time: Geographic Aspects of Communication and Information*, London: Harper Collins Academic, pp.217–31.

Corey, K.E. (1991b), 'An evaluation of Singapore's information technology strategy', paper presented at the International Conference on the Dynamic Transformation of Societies, Seoul, South Korea.

Cronin, F.J., Parker, E.B., Colleran, E.K. and Gold, M.A. (1991), 'Telecommunications infrastructure and economic growth: an analysis of causality', *Telecommunications Policy*, 15, pp.529–35.

Daly, M.T. (1991), 'Transitional economic bases: from the mass production society to the world of finance', in Daniels, P.W. (ed.), *Metropolitan Development: International Perspectives*, London: Routledge, pp.26–43.

Daniels, P.W. (1982), *Service Industries: Growth and Location*, Cambridge: Cambridge University Press.
Daniels, P.W. (1985), *Service Industries: a Geographical Appraisal*, London: Methuen.
Daniels, P.W. (1987), 'Technology and metropolitan office location', *The Service Industries Journal*, 7, pp.274–91.
Daniels, P.W. (1991a), 'Service sector restructuring and metropolitan development: processes and prospects', in Daniels, P.W. (ed.), *Services and Metropolitan Development: International Perspectives*, London: Routledge, pp.1–25.
Daniels, P.W. (1991b), 'Internationalization, telecommunications and metropolitan development: the role of producer services', in Brunn, S.D. and Leinbach, T.R. (eds.), *Collapsing Space and Time: Geographic Aspects of Communication and Information*, London: Harper Collins Academic, pp.149–69.
David, P.A. (1989), *Computer and Dynamo: The Modern Productivity Paradox in a Not-Too-Distant Mirror*, Center for Economic Policy Research Publication 172, Stanford, CA: Stanford University.
Dawson, J. (1979), *The Marketing Environment*, London: Croom Helm.
Dawidziuk, B.M. and Preston, H.F. (1981), 'International communications: network developments and economics', *Telecommunications Journal*, 48, pp.19–31.
de Smidt, M. (1991), 'Spatial diffusion of teleshopping: empirical studies in the Netherlands', *Netcom*, 5, pp.390–400.
Deutsch, K.W. (1956), 'International communication: the media and flows', *Public Opinion Quarterly*, 20, pp.143–60.
Dillman, D.A. (1985), 'The social impacts of information technologies in rural North America', *Rural Sociology*, 50, pp.1–26.
Dordick, H.S. (1990), 'The origins of universal service: history as a determinant of telecommunications policy', *Telecommunications Policy*, 14, pp.223–31.
Downs, A. (1985), 'Living with advanced telecommunications', *Society*, 23 (1), pp.26–34.
Drennan, M.P. (1989), 'Information intensive industries in metropolitan areas of the United States of America', *Environment and Planning A*, 21, pp.1603–618.
Dymond, A. (1987), 'Reducing the number of missing links: regional cooperation and telecommunications development in southern Africa', *Telecommunications Policy*, 12, pp.121–34.
The Economist (1990), 10 March.
The Economist (1991), 5 October.

The Economist (1992), 30 May; 6 June; 20 June.
Ergas, H. and Paterson, P. (1991), 'International telecommunications settlement arrangements: an unsustainable inheritance?', *Telecommunications Policy*, 15, pp.29–48.
Erzberger, H.R. and Sonderegger, U. (1989), 'Satellite offices for computer specialists at Credit Suisse', in Ernste, H. and Jaeger, C. (eds.), *Information Society and Spatial Structure*, London: Belhaven, pp.139–45.
Falk, T. and Abler, R. (1980), 'Intercommunications, distance and geographical theory', *Geografiska Annaler*, 62, pp.59–67.
Fox-Przeworski, J. (1990), 'Information and communication technologies: are there urban policy concerns?', *Netcom*, 4, pp.188–211.
France Information (1984), 121.
Friedman, J. (1986), 'The world city hypothesis', *Development and Change*, 17, pp.69–83.
Friedman, J. and Wolff, G. (1982), 'World city formation: an agenda for research and action', *International Journal for Urban and Regional Research*, 6, pp.309–44.
Gaspar, J. and Jensen-Butler, C. (1990), 'Telecommunications and the location of Portugal in global information space', in Bakis, H. (ed.), *Communications and Territories*, Paris: La Documentation Francaise, pp.165–76.
Genosko, J. (1987), 'The spatial distribution of telematics: modeling and empirical evidence', *Technological Forecasting and Social Change*, 32, pp.281–93.
Gershon, R.A. (1990), 'Global cooperation in an era of deregulation', *Telecommunications Policy*, 14, pp.249–59.
Gillespie, A. (1987), 'Telecommunications and the development of Europe's less-favoured regions', *Geoforum*, 18, pp.229–36.
Gillespie, A. and Robins, K. (1989), 'Geographical inequalities: the spatial bias of the new communications technology', *Journal of Communication*, 39, pp.7–19.
Gillespie, A. and Williams, H. (1988), 'Telecommunications and the reconstruction of regional comparative advantage', *Environment and Planning A*, 20, pp.1311–321.
Glynn, S. (1992), 'Japan's success in telecommunications regulation: a unique regulatory mix', *Telecommunications Policy*, 16, pp.5–12.
Goddard, J.B. (1973), *Office Linkage and Location: A Study of Communication and Spatial Pattern in Central London*, Oxford: Oxford University Press.
Goddard, J.B. (1983), 'The geographical impact of technological

change', in Patten, J. (ed.), *The Expanding City: Essays in Honour of Professor Jean Gottmann*, London: Academic Press, pp.103–24.

Goddard J.B. (1989), 'The city in the global information economy', in Lawton, R. (ed.), *The Rise and Fall of Great Cities: Aspects of Urbanization in the Western World*, London: Belhaven, pp.154–67.

Goddard, J.B. and Gillespie, A.E. (1986), 'Advanced telecommunications and regional economic development', *The Geographical Journal*, 152, pp.383–97.

Goddard, J.B. and Pye, R. (1977), 'Telecommunications and office location', *Regional Studies*, 11, pp.19–30.

Goldmark, P.C. (1972), 'Communication and community', *Scientific American*, 227, pp.143–50.

Goldschmidt, D. (1984), 'Financing telecommunications for rural development', *Telecommunications Policy*, 8, pp.181–203.

Gottmann, J. (1961), *Megalopolis*, New York: The Twentieth Century Fund.

Gottmann, J. (1977), 'Megalopolis and antipolis: the telephone and the structure of the city', in Pool, I.d.S. (ed.), *The Social Impact of the Telephone*, Cambridge, MA: MIT Press, pp.303–17.

Gottmann, J. (1983), *The Coming of the Transactional City*, College Park, MD: University of Maryland Institute for Urban Studies.

Gould, P. (1991), 'Dynamic structures of geographic space', in Brunn, S.D. and Leinbach, T.R. (eds.), *Collapsing Space and Time: Geographic Aspects of Communications and information*, London: Harper Collins Academic, pp.3–30.

Grimes, S. (1992), 'Information technology and regional development: the Irish experience', *Netcom*, 6, pp.281–96.

Gross, D. (1981), 'Space, time and modern culture', *Telos*, 50, pp.59–78.

Gross, D. (1985), 'Temporality and the modern state', *Theory and Society*, 14, pp.53–82.

Hall, P. (1991), 'Moving information: a tale of four technologies', in Brotchie, J., Batty, M., Hall, P., and Newton, P. (eds.), *Cities of the 21st Century: New Technologies and Spatial Systems*, New York: Halsted Press, pp.1–21.

Hamilton, A. (1986), *The Financial Revolution*, Harmondsworth: Viking/Penguin.

Hanneman, G.J. (1986), 'Teleports: an overview', in Lipman, A.D., Sugerman, A.D., and Cushman, R.F. (eds.), *Teleports and the Intelligent City*, Homewood, IL: Dow Jones-Irwin, pp.2–24.

Hansen, S., Cleevely, D., Wadsworth, S., Bailey, H. and Bakewell, O.

(1990), 'Telecommunications in rural Europe: economic implications', *Telecommunications Policy*, 14, pp.207–22.
Hardy, A.P. (1980), 'The role of the telephone in economic development', *Telecommunications Policy*, 4, pp.278–86.
Harkness, R.C. (1973), 'Communication innovations, urban form and travel demand: some hypotheses and a bibliography', *Transportation*, 2, pp.153–93.
Hart, J.A. (1988), 'The politics of global competition in the telecommunications industry', *The Information Society*, 5, pp.169–201.
Heng, T.M. and Low, L. (1990), 'Towards greater competition in Singapore's telecommunications', *Telecommunications Policy*, 14, pp.303–14.
Henry, W.A. III. (1992), 'History as it happens', *Time*, 6 January.
Hepworth, M. (1990), *Geography of the Information Economy*, London: Belhaven.
Hepworth, M. (1991a), 'Information cities in Europe 1992', *Telecommunications Policy*, 15, pp.175–81.
Hepworth, M. (1991b), 'Information technology and the global restructuring of capital markets', in Brunn, S.D. and Leinbach, T.R. (eds.), *Collapsing Space and Time: Geographic Aspects of Communication and Information*, London: Harper Collins Academic, pp.132–48.
Hepworth, M. and Ducatel, K. (1992), *Transport in the Information Age: Wheels and Wires*, London: Belhaven Press.
Hoffman, D. (1991), 'Global communications network was pivotal in defeat of junta', *The Washington Post*, 23 August, A27.
Hottes, K. (1992), 'Submarine cables in our times: a report on competition between seacables and satellites', paper presented at the IGU pre-congress symposium, Washington, DC.
Howland, M. (1991), 'Producer services: will they follow manufacturing out of urban centers?', *Economic Development Commentary*, 15 (3), pp.4–8.
Howland, M. (1992), 'Technological change and the spatial restructuring of data entry and processing services', *Technological Forecasting and Social Change*, 37 (forthcoming).
Hudson, H.E. (1984), *When Telephones Reach the Village: the Role of Telecommunications in Rural Development*, Norwood, NJ: Ablex Publishing.
Hudson, H.E. (1991), 'Telecommunications in Africa: the role of the ITU', *Telecommunications Policy*, 15, pp.343–50.
Innis, H.A. (1950), *Empire and Communications*, Oxford: Clarendon Press.

References

Innis, H.A. (1951), *The Bias of Communications*, Toronto: University of Toronto Press.

International Encyclopedia of Communications (1989), New York: Oxford University Press.

International Telecommunications Union (ITU) (1983), *Telecommunications for Development*, Geneva: ITU.

Janelle, D.G. (1968), 'Central place development in a time-space framework', *The Professional Geographer*, 20, pp.5–10.

Janelle, D.G. (1991), 'Global interdependence and its consequences', in Brunn, S.D. and Leinbach, T.R. (eds.), *Collapsing Space and Time: Geographic Aspects of Communication and Information*, London: Harper Collins Academic, pp.49–81.

Jipp, A. (1963), 'Wealth of nations and telephone density', *Telecommunications Journal*, pp.199–201.

Jowett, P. and Rothwell, M. (1986), *The Economics of Information Technology*, London: Macmillan, 1986.

Jussawalla, M. and Cheah, C.W. (1987), *The Calculus of International Communications: A Study in the Political Economy of Transborder Data Flows*, Littleton, CO: Libraries Unlimited.

Jussawalla, M. and Ogden, M.R. (1989), 'The Pacific Islands: policy options for telecommunications investment', *Telecommunications Policy*, 13, pp.40–50.

Keen, P.G.W. (1988), *Competing in Time: Using Telecommunications for Competitive Advantage*, New York: Ballinger.

Kellerman, A. (1984), 'Telecommunications and the geography of metropolitan areas', *Progress in Human Geography*, 8, pp.222–46.

Kellerman, A. (1985), 'The evolution of service economies: a geographical perspective', *The Professional Geographer*, 37, pp.133–43.

Kellerman, A. (1986a), 'The diffusion of BITNET: a communications system for universities', *Telecommunications Policy*, 10, pp.88–92.

Kellerman, A. (1986b), 'Telecommunicated universities', *Land Use Policy*, 3, pp.213–20.

Kellerman, A. (1989), *Time, Space, and Society: Geographical Societal Perspectives*, Dordrecht: Kluwer.

Kellerman, A. (1990), 'International telecommunications around the world: a flow analysis', *Telecommunications Policy*, 14, pp.461–75.

Kellerman, A. (1991a), 'The decycling of time and the reorganization of urban space', *Cultural Dynamics*, 4, pp.38–54.

Kellerman, A. (1991b), 'The role of telecommunications in assisting peripherally located countries: the case of Israel', in Brunn, S.D. and Leinbach, T.R. (eds.), *Collapsing Space and Time: Geographic*

Aspects of Communication and Information, London: Harper Collins Academic, pp.252–77.

Kellerman, A. (1992a), *US International Telecommunications 1961–1989: Temporal and Spatial Aspects by Various Modes of Measurement*, International Center for Telecommunications Management (ICTM) Research Paper 8, Omaha: University of Nebraska.

Kellerman, A. (1992b), 'US international telecommunications 1961–1988: an international movement model', *Telecommunications Policy*, 16, pp.404–14.

Kellerman, A. and Cohen, A. (1992), 'International telecommunications as international movement: the case of Israel, 1951–1988', *Telecommunications Policy*, 16, pp.156–66.

Kellerman, A. and Krakover, S. (1986), 'Multi-sectoral urban growth in space and time: an empirical approach', *Regional Studies*, 20, pp.117–29.

Kern, S. (1983), *The Culture of Time and Space 1880–1918*, Cambridge, MA: Harvard University Press.

King, A.D. (1990), *Global Cities: Post-Imperialism and the Internationalization of London*, London: Routledge.

Klaassen, L.H., Wagenaar, S. and van der Weg, A. (1972), 'Measuring psychological distance between the Flemings and the Waloons', *Papers of the Regional Science Association*, 29, pp.45–62.

Knight, R.V. and Gappert, G. (1984), 'Cities and the challenge of the global economy', in Bingham, R.D. and Blair, J.P. (eds.), *Urban Economic Development*, Urban Affairs Annual Reviews, 27, Beverly Hills, CA: Sage, pp.63–78.

Koppelman, F., Salomon, I. and Proussaloglou, K. (1991), 'Teleshopping or store shopping? A choice model for forecasting the use of new telecommunications-based services', *Environment and Planning B: Planning and Design*, 18, pp.473–89.

Kraut, R.E. (1989), 'Telecommuting: the trade-offs of home work', *Journal of Communication*, 39, pp.19–47.

Kumar, A. (1990), 'Impact of technological developments on urban form and travel behaviour', *Regional Studies*, 24, pp.137–48.

Kutay, A. (1986), 'Effects of telecommunications technology on office location', *Urban Geography*, 7, pp.243–57.

Kutay, A. (1988), 'Technological change and spatial transformation in an information economy: 2. The influence of new information technology on the urban system', *Environment and Planning A*, 20, pp.707–18.

Langdale, J.V. (1978), 'The growth of long-distance telephony in the Bell system: 1875–1907', *Journal of Historical Geography*, 4, pp.145–59.

Langdale, J.V. (1983), 'Competition in the United States' long-distance telecommunications industry', *Regional Studies*, 17, pp.393–409.
Langdale, J.V. (1985), 'Electronic funds transfer and the internationalisation of the banking and finance industry', *Geoforum*, 16, pp.1–13.
Langdale, J.V. (1987), 'Telecommunications and electronic information services in Australia', in Brotchie, J.F., Hall, P. and Newton, P.W. (eds.), *The Spatial Impact of Technological Change*, London: Croom Helm, pp.89–103.
Langdale, J.V. (1989a), 'The geography of international business telecommunications: the role of leased networks', *Annals of the Association of American Geographers*, 79, pp.501–22.
Langdale, J.V. (1989b), 'International telecommunications and trade in services: policy perspectives', *Telecommunications Policy*, 13, pp.203–21.
Langdale, J.V. (1989c), 'Telecommunications and the multi-function polis: international, urban and regional development implications', *Netcom*, 3, pp.12–26.
Langdale, J.V. (1991a), 'Restructuring telecommunications', *Australian Geographer*, 22, pp.124–26.
Langdale, J.V. (1991b), 'Telecommunications and international transactions in information services', in Brunn, S.D. and Leinbach, T.R. (eds.), *Collapsing Space and Time: Geographic Aspects of Communication and Information*, London: Harper Collins Academic, pp.193–214.
Langdale, J.V. (1992), 'Transnational corporations and the adoption of information and communications technologies in the Asia-Pacific region', *Netcom*, 6, pp.195–217.
LaRose, R. and Mettler, J. (1989), 'Who uses information technologies in rural America?', *Journal of Communication*, 39, pp.48–60.
Lauder, G. (1990), 'Telecommunications and regional development in the European community: the STAR programme', in Bakis, H. (ed.), *Communications and Territories*, Paris: La Documentation Francaise, pp.289–95.
Leff, N.H. (1983), 'Externalities, information costs, and social development: an example from telecommunications', *Economic Development and Cultural Change*, 32, pp.255–76.
Lehman-Wilzig, S. (1981), 'Will cities become obsolete?', *Telecommunications Policy*, 5, pp.326–28.
Lerner, N.C. (1968), *Evaluation and Development of Forecasting Techniques for US International Telecommunications Traffic*, PhD dissertation, Washington, DC: American University.

Lesko, A.P. (1989), 'The growth of telephone service in Argentina', *Netcom*, 3, pp.423-35.

Lesko, A.P. (1990), 'On the geography of telecommunications: telephone development, diffusion and traffic; the case of Tucuman, Argentina', unpublished MA thesis, Department of Social and Economic Geography, University of Lund.

Lesko, A.P. (1992), 'Telephone flows at the provincial and national level: an example from Argentina', paper presented at the IGU pre-congress telecommunications symposium, Washington, DC.

Lesser, B. and Hall, P. (1987), *Telecommunications Services and Regional Development: the Case of Atlantic Canada*, Halifax, Nova Scotia: The Institute for Research on Public Policy.

Lewis, L.T. (1989), 'BITNET: a tool for communications among geographers', *The Professional Geographer*, 41, pp.470-79.

Lewis, N.D. and Mukaida, L.V.D. (1991), 'Telecommunications in the Pacific region: the PEACESAT experiment', in Brunn, S.D. and Leinbach, T.R. (eds.), *Collapsing Space and Time: Geographic Aspects of Communication and Information*, London: Harper Collins Academic, pp.232-51.

Long, M. (1992), *World Satellite Almanac: The Complete Guide to Satellite Transmission Technology*, Third edition, Indianapolis, IN: Howard D. Sams.

Lord, J.D. (1992), 'Geographic deregulation of the US banking industry and spatial transfers of corporate control', *Urban Geography*, 13, pp.25-48.

Lyon, D. (1986), 'From "post-industrialism" to "information society": a new social transformation?', *Sociology*, 20, pp.577-88.

Ma'ariv (1991), 2 December (Hebrew).

Ma'ariv (1992a), 8 January and 19 January (Hebrew).

Ma'ariv (1992b), 10 February and 17 February (Hebrew).

Ma'ariv (1992c), 29 April (Hebrew).

MacMahon, A.M. (1980), 'Computer communication-concepts and technology', in House, W.C. (ed.), *Electronic Communications Systems*, New York: Petrocelli Books, pp.13-21.

McCarroll, T. (1991), 'What new age?', *Time*, 12 August.

McLuhan, M. (1964), *Understanding Media*, New York: McGraw-Hill.

Mandeville, T. (1983), 'The spatial effects of information technology', *Futures*, 15, pp.65-72.

Martin, J. (1991), 'Global business', *Pan Am Clipper*, 31 (4), pp.13.

Martin, M. (1991), 'Communication and social form: the development of the telephone, 1876-1920', *Antipode*, 23, pp.307-33.

Marvin, C. (1988), *When Old Technologies Were New: Thinking about Electric Communication in the Late Nineteenth Century*, New York: Oxford University Press.

Mayo, J.K., Heald, G.R. and Klees, S.J. (1992), 'Commercial satellite telecommunications and national development: lessons from Peru', *Telecommunications Policy*, 16, pp.67-79.

Meyrowitz, J. (1985), *No Sense of Place: The Impact of Electronic Media on Social Behavior*, New York: Oxford University Press.

Mintz, L. (1992), 'Satellite network choice due', *Washington Post*, 5 August.

Morgan, K. and Pitt, D. (1989), 'Coping with turbulence: corporate strategy, regulatory regimes, innovation and the new telecommunication paradigm', in Granham, N. (ed.), *European Telecommunications Policy Research*, Amsterdam: IOS, pp.19-39.

Morgan, M. and Birley, P.D. (1990), 'Multifunction polis: implications for the telecommunications infrastructure', in Bakis, H. (ed.), *Communications and Territories*, Paris: La Ducomentation Francaise, pp.241-52.

Moss, M.L. (1986a), 'Telecommunications and the future of cities', *Land Development Studies*, 3, pp.33-44.

Moss, M.L. (1986b), 'Telecommunications systems and large world cities: a case study of New York', in Lipman, A.D., Sugerman, A.D., and Cushman, R.F. (eds.), *Teleports and the Intelligent City*, Homewood, IL: Dow Jones-Irwin, pp.378-97.

Moss, M.L. (1987a), 'Telecommunications and international financial centres', in Brotchie, J.F., Hall, P. and Newton, P.W. (eds.), *The Spatial Impact of Technological Change*, London: Croom Helm, pp.75-88.

Moss, M.L. (1987b), 'Telecommunications, world cities, and urban policy', *Urban Studies*, 24, pp.534-46.

Moss, M.L. (1989), 'The information city in the global economy', paper presented at the Third International Workshop on Innovation, Technological Change and Spatial Impacts, Cambridge.

Moss, M.L. (1991), 'The new fibers of urban economic development', *Portfolio*, 4, 11-18.

Moss, M.L. and Brion, J.G. (1991), 'Foreign banks, telecommunications, and the central city', in Daniels, P.W. (ed.), *Services and Metropolitan Development: International Perspectives*, London: Routledge, pp.265-84.

Moss, M.L. and Donau, A. (1986), 'Offices, information technology, and locational trends', in Black, J.T., Roark, K.S., and Schwartz, L.S. (eds.), *The Changing Office Workplace*, Washington, DC:

The Urban Land Institute, pp.171–82.

Mowlana, H. and Wilson, L.J. (1990), *The Passing of Modernity: Communication and the Transformation of Society*, New York: Longman.

Nicol, L.Y. (1983), *Communication, Economic Development and Spatial Structures: Theoretical Framework*, Working paper 405, Berkeley, CA: University of California, Institute of Urban and Regional Development.

Nicol, L. (1985), 'Communications technology: economic and spatial impacts', in Castells, M. (ed.), *High Technology, Space, and Society*, Urban Affairs Annual Reviews 28, Beverly Hills, CA: Sage, pp.191–209.

Nijkamp, P., Rietveld, P., and Salomon, I. (1990), 'Barriers in spatial interactions and communication: a conceptual exploration', *Annals of Regional Science*, 24, pp.237–52.

Nijkamp, P. and Salomon, I. (1989), 'Future spatial impacts of telecommunications', *Transportation Planning and Technology*, 13, pp.275–87.

Nilles, J.M., Carlson, F.R. Jr., Gray, P., and Hanneman, G.J. (1976), *The Telecommunications-Transportation Tradeoff: Options for Tomorrow*, New York: Wiley-Interscience.

Nora, S. and Minc, A. (1980), *The Computerization of Society*, Cambridge, MA: MIT Press.

Noyelle, T. and Peace, P. (1991), 'Information industries: New York's new export base', in Daniels, P.W. (ed.), *Services and Metropolitan Development: International Perspectives*, London: Routledge, pp.285–304.

Olson, M.H. (1989), 'Telework: effects of changing work patterns in space and time', in Ernste, H. and Jaeger, C. (eds.), *Information Society and Spatial Structure*, London: Belhaven Press, pp.129–37.

O'Neill, H.O. and Moss, M.L. (1991), *Reinventing New York: Competing in the Next Century's Global Economy*, New York: Urban Research Center, New York University.

Ono, R. (1990), 'Improving development assistance for telecommunications', *Telecommunications Policy*, 14, pp.476–87.

Organization for Economic Cooperation and Development (OECD) (1975), *Applications of Computer/Telecommunications Systems*, Information Studies 8.

Organization for Economic Cooperation and Development (OECD) (1985), *Update of Information Sector Statistics*, Committee for Information, Computer and Communications Policy.

Parker, E.B., Hudson, H.E., Dillman, D.A., and Roscoe, A.D. (1989), *Rural America in the Information Age: Telecommunications Policy for Rural Development*, Lanham, MD: The Aspen Institute and the University Press of America.
Perlmutter, H. (1979), 'Philadelphia: the emerging international city', Philadelphia: La Salle College.
Phillips, A. (1991), 'Changing markets and institutional inertia: a review of US telecommunications policy', *Telecommunications Policy*, 15, pp.49–61.
Phillips, K.A. (1986), 'Teleports: what's it all about', in Lipman, A.D., Sugerman, A.D., and Cushman, R.F. (eds.), *Teleports and the Intelligent City*, Homewood, IL: Dow Jones-Irwin, pp.82–94.
Pool, I.d.S. (ed.) (1977), *The Social Impact of the Telephone*, Cambridge, MA: MIT Press.
Pool, I.d.S. (1990), *Technologies without Boundaries: On Telecommunications in a Global Age*, Cambridge, MA: Harvard University Press.
Pozo de Bisceglia, S.I. (1953), *Geografia de Telecommunicaciones*, Buenos Aires: Arbo Editores.
Pred, A. (1973), 'The growth and development of systems of cities in advanced economies', in Pred, A. and Tornqvist, G. (eds.), *Systems of Cities and Information Flows*, Lund: University of Lund, Lund Studies in Geography Series B38, pp.9–82.
Pred, A. (1977), *City Systems in Advanced Economies*, London: Hutchinson.
Price, D.G. and Blair, A.M. (1989), *The Changing Geography of the Service Sector*, London: Belhaven Press.
Pye, R. and Lauder, G. (1987), 'Regional aid for telecommunications in Europe', *Telecommunications Policy*, 11, pp.99–113.
Quarterman, J.S. (1990), *The Matrix: Computer Networks and Conferencing Systems Worldwide*, Bedford, MA: Digital Press.
Qvortrup, L. (1990), 'The Nordic telecottages: community teleservice centres for rural regions', *Telecommunications Policy*, 13, pp.59–68.
Ratzel, F. (1897), *Politische Geografie*, Munich: R. Oldenbourg.
Raulet, G. (1991), 'The new utopia: communication technologies', *Telos*, 87, pp.39–58.
Rietveld, P. and Janssen, L. (1990), 'Telephone calls and spatial interactions: the case of the Netherlands', *Netcom*, 4, pp.132–44.
Rietveld, P. and Rossera, F. (1992), 'Telecommunications demand: the role of barriers', paper prepared for the IGU pre-congress telecommunications symposium, Washington, DC.
Rietveld, P., Rossera, F., and van Nierop, J. (1992), 'Technology

substitution and diffusion in telecommunications: the case of telex and telephone network', paper prepared for the IGU pre-congress telecommunications symposium, Washington, DC.

Robins, K. and Hepworth, M. (1988), 'Electronic spaces: new technologies and the future of cities', *Futures*, 20 (2), pp.155–76.

Robinson, A.L. (1977), 'Impact of electronics on employment: productivity and displacement effects', *Science*, 195, pp.1179–184.

Robinson, F. (1984), 'Regional implications of information technology', *Cities*, 1, pp.356–61.

Robinson, P. (1991), 'The international dimension of telecommunications policy issues', *Telecommunications Policy*, 15, pp.95–100.

Rossera, F. (1990), 'Discontinuities in communications among communities of different language in Switzerland: an analysis based on data of long-distance telephone calls', *Netcom*, 4, pp.119–31.

Rubinstein, A. (1991), 'The undocumented conversations', *Ha'aretz*, 4 April (Hebrew).

Salomon, I. (1985), 'Telecommunications and travel: substitution or modified mobility?', *Journal of Transport Economics and Policy*, 19, pp.219–35.

Salomon, I. (1988), 'Transportation–telecommunication relationship and regional development', in Giaoutzt, M. and Nijkamp, P. (eds.), *Informatics and Regional Development*, Aldershot: Avebury (Gower), pp.90–102.

Salomon, I. and Razin, E. (1988), 'The geography of telecommunications systems: the case of Israel's telephone system', *Tijdschrift voor Economische en Sociale Geografie*, 79, pp.122–34.

Salomon, I. and Salomon, M. (1984), 'Telecommuting – the employees' perspective', *Technological Forecasting and Social Change*, 25, pp.15–28.

Salomon, I., Schneider, H.N., and Schofer, J. (1991), 'Is telecommunicating cheaper than travel? An examination of interaction costs in a business setting', *Transportation*, 18, pp.291–318.

Sassen-Koob, S. (1985), 'Capital mobility and labor migration: their expression in core cities', in Iimberlake, M. (ed.), *Urbanization in the World System*, New York: Academic Press, pp.231–65.

Sassen, S. (1991), *The Global City: New York, London, Tokyo*, Princeton, NJ: Princeton University Press.

Saunders, R.J., Warford, J.J., and Wellenius, B. (1983), *Telecommunications and Economic Development*, Baltimore: The Johns Hopkins University Press.

Sawhney, H. (1992), 'Rural telephone companies: diverse outlooks and shared concerns', *Telecommunications Policy*, 16, pp.16–26.

Schmidt, S.K. (1991), 'Taking the long road to liberalization: telecommunications reform in the Federal Republic of Germany', *Telecommunications Policy*, 15, pp.209–22.

Schwartz, A. (1992), 'The geography of corporate services: a case study of the New York urban region', *Urban Geography*, 13, pp.1–24.

Scott, A.J. (1982), 'Locational patterns and dynamics of industrial activity in the modern metropolis', *Urban Studies*, 19, pp.111–42.

Sheffer, D. (1988), 'The effect of various means of communication on the operation and location of high-technology industries', in Giaoutzi, M. and Nijkamp, P. (eds.), *Informatics and Regional Development*, Aldershot: Avebury (Gower), pp.166–81.

Short, J., Williams, E. and Christie, B. (1976), *The Social Psychology of Telecommunications*, London: John Wiley.

Siemens (1991), *1991 International Telecommunications Statistics*, Munich: Siemens.

Simpson, A. (1984), *Facsimile*, London: Eurodata Foundation.

Singlemann, J. (1978), *From Agriculture to Services: the Transformation of Industrial Employment*, Beverly Hills, CA: Sage.

Smith, K.A. and Healy, P.E. (1987), 'Transborder data flows: the transfer of medical and other scientific information by the United States', *The Information Society*, 5, pp.67–76.

Snow, M.S. (1988), 'Telecommunications literature: a critical review of the economic, technological and public policy issues', *Telecommunications Policy*, 12, pp.153–83.

Sproull, L. and Kiesler, S. (1991), *Connections: New Ways of Working in the Networked Organization*, Cambridge, MA: MIT Press.

Stanley, K.B. (1991), 'Balance of payments, deficits, and subsidies in international communications services: a new challenge to regulation', *Administrative Law Review*, 43, pp.411–38.

Staple, G.C. (ed.), (1991), *The Global Telecommunications Traffic Report – 1991*, London: International Institute of Communications.

Staple, G.C. and Mullins, M. (1989a), 'Telecom traffic statistics – MiTT matter: improving economic forecasting and regulatory policy', *Telecommunications Policy*, 13, pp.105–28.

Staple, G.C. and Mullins, M. (1989b), *Global Telecommunication Traffic Flows and Market Structures: A Quantitative Review*, London: International Institute of Communications.

Taylor, L.D. (1983), 'Problems and issues in modelling telecommunications demand', in Courville, L., de Fontenay, A., and Dobell, R. (eds.), *Economic Analysis of Telecommunications*, Amsterdam: Elsevier Science Publishers BV (North Holland), pp.181–98.

Television Factbook 1981/82 (1982), Washington, DC: Television Digest.

Thrift, N. (1987), 'The fixers: the urban geography of international commercial capital', in Henderson, J. and Castells, M. (eds.), *Global Restructuring and Territorial Development*, London: Sage, pp.203–33.

Toffler, A. (1981), *The Third Wave*, New York: Bantam Books.

Toffler, A. (1990), *Power Shift*, New York: Bantam Books.

Toong, H.m.D. and Gupta, A, 'Personal computers', *Scientific American*, 247, pp.86–107.

Tornqvist, G. (1968), 'Flows of information and the location of economic activities', *Geografiska Annaler*, 50B, pp.99–107.

Ullman, E.L. (1957), *American Commodity Flow*, Seattle, WA: University of Washington Press.

US Bureau of the Census (1991), *Statistical Abstract of the US 1991*, Washington, DC: US Government Printing Office.

US Federal Communications Commission (FCC) (1961–1990), *Statistics of Communications Common Carriers*, Washington, DC: US Government Printing Office.

US Federal Communications Commission (FCC) (1988), *International Accounting Rates and the Balance of Payments Deficit in Telecommunication Services*, (mimeo).

Ure, J. (1989), 'The future of telecommunications in Hong Kong', *Telecommunications Policy*, 13, pp.371–78.

Warf, B. (1989), 'Telecommunications and the globalization of financial services', *The Professional Geographer*, 41, pp.257–71.

Warf, B. (1991), 'The internationalization of New York services', in Daniels, P.W. (ed.), *Services and Metropolitan Development: International Perspectives*, London: Routledge, pp.245–64.

The Washington Post (1991), 29 August.

Webber, M.M. (1968), 'The post-city age', *Daedalus*, 97, pp.1091–110.

Wilson, A. (1991), *Interavia Space Directory 1991–92*, London: Jane's Information Group.

Wilson, M.I. (1992), 'The office farther back: business services, productivity, and the offshore back office', unpublished paper.

Wise, A. (1971), 'The impact of electronic communications on metropolitan form', *Ekistics*, 188, pp.22–31.

World Space Industry Survey: Ten Year Outlook 1991/1992 Edition (1991), Paris: Euroconsult.

The World's Telephones 1977–1988, Sarasota, FL: AT&T.

Glossary

ANALOGUE TRANSMISSION:
In this transmission technology information is represented by a continuously variable signal.

COMMON CARRIER:
A company that carries electronic signals for voice, text or data from place to place for a fee paid by customers at large.

DIGITAL TRANSMISSION:
In this transmission technology information is encoded by discrete amounts of ones and zeros.

FCC:
The US Federal Communications Commission. Established in 1934, it regulates and controls telecommunications and broadcasting services.

INFORMATION TECHNOLOGY (IT):
A term normally used to jointly describe telecommunications and computers.

ISDN:
Integrated Services Digital Network, combines voice, data, fax, and video transmission in the same channels.

LANS:
Local Area Networks are computer networks within a closed telecommunications system.

LATA:
Local Access Transport Area, or the geographical area within which a Bell operating company may offer long-distance services in the US.

PACKET SWITCHES:
Digital telephone exchanges in which messages are fragmented into 'packets' consisting of a definite amount of characters.

POTS:
Plain Old Telephone Service, or voice telephony.

PTT:
Post, Telegraph, and Telephone. The traditional governmental department dealing with these services.

RBOC:
Regional Bell Operating Company, or 'baby Bell', one of the seven regional telephone companies in the US, formed with the divestiture of AT&T.

VANS:
Value Added Networks are information services provided over voice telephone, or through computer or fax terminals.

Index

AT&T, 27, 28, 46, 50, 52, 63, 75, 77, 105, 144–5, 153, 165, 167, 169, 189
Abbatiello, J., 164, 198
Abler, R.F., xvi, 3, 4, 10, 11, 14, 15, 16, 20, 21, 33, 36, 54, 60, 65, 70–1, 74, 77–8, 95, 96, 108, 117, 122, 126, 198–9, 127, 145, 196, 203
Access, 30, 32, 34, 55, 60, 88–9, 126, 138, 192
Accessibility, 30, 194, 195
Ackerman, E.A., 15, 199
Adams, J.S., 10, 11, 74, 198
Adrian-Bueckling, M., 41, 199
Africa, 48, 131
  African, 131, 134, 157, 176
Agriculture, 122, 124, 139
Airports, 13, 104, 151
Akhavan-Majid, R., 142, 146, 199
Akwule, R.U., 142, 199
Albany, 77
America, 27, 105, 169
  rural, 123, 125–7, 182
  urban, 126
American, 25, 27, 41, 50, 63, 68, 101, 103, 104, 109, 125, 127, 144, 145, 147, 161, 166, 173, 178–9, 181, 182, 183, 190–2, 196
  cities, 84
  economy, 122, 191
  households, xv, 76
  market, 28, 38, 83
  society, 75, 190–1
  universities, 5, 184
Americans, 41, 68, 127, 165
Amsterdam, 55, 93, 184
Analogue, 25, 26, 53; see also switches
Answering
  machines, 32, 84
  service, 27, 54

Antenna, 39, 48, 54, 81, 165; see also microwaves
Antipolitan approach, 95
Antonelli, C., 121, 199
Arab, 72, 177
Argentina, xvi, 128, 143
Aronson, S., 32, 33, 199
Asia, 182, 183
  Asian, 52, 66, 134, 176
ASIANET, 184
Atlanta, 114
Atlantic, 40, 48, 50, 51–2, 66
  Atlantic region, 66
  transatlantic, 47, 50
Australia, 46, 150, 166–7, 181
  Australian, 115, 166
Austria, 174
  Austrians, 70

Bahrain, 100
Bailey, H., 204–5
Bakewell, O., 204–5
Bakis, H., xvi, 12, 55, 180, 199
Ball, D.W., 30, 32, 33, 199
Baltimore, 99
Bank, 35, 38, 101, 104–6, 109, 112, 180, 181–2
  offices, 106, 182–3
Banking, 17, 101, 104, 106, 114, 123
  foreign, 105, 109
  international, 101, 102, 104, 106, 109, 151; see also regional
Barbados, 182–3
Barcelona, 93
Barriers, 61, 68–72, 88, 108, 153, 162, 194; see also flow
Bas Normandie, 124
Batty, M., 76, 119, 184, 199
Bavaria, 121

Beesley, M.E., 146, 199
Belgium, 70
Bell (Telephone Company), 21, 70, 77, 105, 126, 127, 144-5
Bell, D., 4, 6-7, 9, 10, 200
Beniger, J.R., 6, 22, 76, 200
Berry, B.J.L., 75, 200
Beyers, W.B., 120, 122, 123, 127, 200
Bezek, xvii, 63, 64, 65, 67, 142, 148, 200; see also Israeli
Bikson, T.K., 12, 200
Birley, P.D., 115, 210
BITNET, 75, 80, 184
Blair, A.M., 80, 85, 212
Boston, 77, 114
Brazil, 181
Brion, J.G., 105, 109, 210
Britain, 38, 173, 181, 184
 British, 5, 25, 27, 28, 38, 103, 106, 146, 147, 150, 165, 179, 183, 184, 196
British Telecom, 28, 146, 184
Brooker-Gross, S.R., 74-5, 77, 83, 85, 200
BRT (British Rail Telecommunications), 38
Brunn, S.D., xvi, 130, 200
Brussels, 101, 155
Buenos Aires, 128
Buffalo, 77
Bulgaria, 177
Business, 5, 20, 27, 56, 58, 63, 66, 68, 69, 79, 80, 89, 96, 99, 103, 108, 109, 112, 114, 117, 119, 125, 139, 140, 143, 145, 146, 148, 151, 152, 153, 160, 161, 162, 167, 181, 185, 188, 191, 194, 195, 196
 calls, 61, 68, 85-6
 community, 77, 80, 103
 meetings, 70, 83
 sector, 94, 129
 visits, 178; see also services, travel, information, tourism, rural, international
Buyer, M. 109, 200

C-band, 41
Cable, xv, 47, 48, 50, 52, 81
 systems, 39, 48, 54; see also networks
Cable & Wireless, 53, 165
Cable TV, 23, 25, 36, 73, 75, 81-3, 84, 195, 196
Cables, 4, 13, 24, 25, 26, 30, 36, 38, 48, 51, 52, 63, 81, 131, 141, 155, 165
 coaxial, 36, 48

maritime, 13, 20, 36, 40, 46, 47-53, 165
Call minutes, 153, 165, 168, 180
Cambridge, MA, 184
Canada, 65, 81, 125, 127, 140, 144, 164, 166, 169, 172, 181, 184
Canadian, 183
Capital, xvii, 9, 17, 63, 99, 101, 102, 104, 105, 106, 114, 133, 138, 140, 141, 143, 148, 153, 155, 157, 160, 177-8, 188, 194
 accumulation, 118-9, 194
 exchanges, 179, 192
 markets, 35, 93, 97, 99, 104, 105, 109, 121, 155, 188, 193; see also global, international
Carey, J., 27, 74, 76, 77, 83, 200
Carlson, F.R. Jr., 211
Caribbean, 65, 182-3
Carriers; see common carriers
Case, D.O., 142, 200
Castells, M., 14, 101, 105, 106, 182-3, 190, 194, 200
Cellular telephones, 23, 27, 29-30, 32, 38, 129, 131, 141, 143, 147, 194, 195
Centre for Telecommunications Development (CTD), 164
Central business districts (CBDs), 15, 61, 95, 108, 111, 113, 114, 115, 187, 193
 specialization, 113-4; see also downtown
Charlotte, 106
Cheah, C.W., 180-1, 206 Cheong, K., 166-7, 201
Cherry, C., 5, 201
Chicago, 38, 77, 86, 109
China, 169, 183
Christaller, W., 14, 93, 201
Christie, B., 85, 201, 214
Cities, 61, 75, 83, 93-115, 140, 150, 193
 domestic, 98, 99
 global, 114
 large, 74, 75, 89, 94, 95, 98, 109-11, 115, 139, 182, 188
 leading, 103, 194
 major, 71, 74, 118-9, 150, 188, 194
 provincial, 120
 small, 94, 109, 111
 world, 61, 98, 99-102, 113, 190; see also core
City, 15, 54, 73, 80, 83, 93-7, 100, 104, 105, 113, 114, 128, 151, 188
 -wide, 56
 global, 99
 inner-, 111

inter-, 77, 88
international, 99
transactional, 98
world, 48, 99–100
City-country, 148
City-states, 101
Clapp, J.M., 128–9, 201
Clark Circle, 40
Cleevely, D., 204–5
Cleveland, 77
CNN, 157
Codding, G.A. Jr., 163, 201
Cohen, A., 61, 73, 159–60, 172, 174, 178–9, 207
Cohen, R.B., 99, 100, 201
Colleran, E.K., 201
Cologne, 93
Colonial,
  imprints, 176–7
  powers, 168
Commodities, xvii, 12, 17, 63, 73, 99, 102, 108, 113, 153, 155, 157, 160–1, 177–9, 189, 194; see also goods
Common carriers, 26, 53, 146, 150, 165, 178; see also international, Israeli
Commuting, 83, 112, 123
Complementarity, 161
Compunications, 4
Computer, xv, 4, 12, 20, 21, 25, 34, 37, 38, 54, 61, 84, 108, 121, 123 162, 167, 183
  business, 27, 145
  companies, 55, 114
  services, 5, 101, 195, 196
  systems, 5, 12, 37, 181
  terminals, 25, 27; see also network
Computerization, 72
Computers, 4–5, 8, 13, 21, 27, 28, 37, 53, 73, 75–6, 80–1, 84, 145, 151, 183, 196
Computing, 4, 73
COMSAT company, 41
Concentration, 17, 35, 54, 55, 56, 94–7, 98, 104, 108, 111, 115, 187–90; see also urban
Confravision, 28; see also teleconferencing, videoconferencing
Copenhagen, 100
Core, 113, 128, 129, 176, 190
  areas, 124, 128
  cities, 120
  countries, 133, 176; see also cities
Cores, 34, 35, 58, 66, 89, 117–20, 127, 128, 131, 169, 189
  economic, 34

urban, 34, 118; see also world
Corey, K.E., 98, 142, 201
Corporations, 12, 38; see also transnational corporations
Cronin, F.J., 135, 136, 201
Cross-subsidization, 71, 110, 126, 146
Cuba, 176
Cultural, 100, 129, 168, 169, 172, 176, 180, 196
  patterns, 169, 174–6; see also barriers
Culture, 5, 100, 175

Dallas, 114
Daly, M.T., 76, 201
Daniels, P.W., 7, 9, 102–4, 106, 108–10, 111, 131
Data, 3, 5, 13, 21, 25, 36, 38, 54, 60, 61, 63, 123, 131, 153, 160, 161, 179, 181, 189, 195
  processing, 5, 37, 93, 111, 112, 115, 129, 146, 181, 182–3; see also digital, transmission
David, P.A., 196, 202
Dawson, J., 7, 202
Dawidziuk, B.M., 22, 202
de Smidt, M., 84–5, 202
Decentralization, 33, 34, 89, 94, 111, 113, 122, 188, 190; see also office
Delocalization, 33, 34, 89, 195
Demand, 7, 8, 30, 38, 40, 41, 47, 48, 50, 53, 56, 64, 74, 88–9, 95, 109, 110, 114, 120, 123, 125, 138, 141, 145, 148, 149, 150, 157, 162, 166, 167, 178, 181, 188, 192, 194
Denmark, 175
Deregulation, 70, 106, 109–10, 143, 145–6, 150, 193; see also regulation
Detroit, 105
Deutsch, K.W., 64, 202
Deutsches Bundespost (DBP), 147
Developed countries, 63, 72, 124, 128, 136, 143
Developing countries, see developing nations
Developing nations, 41, 76, 131, 135, 163, 167, 176; see also less developed countries
Development, 4, 7, 12, 17, 21–6, 35, 57, 59, 61, 63, 73, 79, 88, 89, 93, 95, 97, 101, 104, 106–10, 111, 114, 116–29, 131, 133, 135–8, 144, 145, 163, 164, 167, 181, 182, 187, 190, 191, 194, 196
  process, 116, 120, 138; see also economic, regional, international

Index

Diffusion, xvi, 30, 60, 80, 82, 88, 184, 190, 193
  of innovations, xvii, 17, 73–83
  process, 73, 74–5, 94
  trends, 75–6
Digital, 25, 26, 38, 75, 126, 148
  equipment, 25, 53, 190
  exchange service, 26, 148
  information, 26, 35, 180
  switching, 4, 23, 26, 37, 72, 155
  technology, 79, 125; see also telephone, transmission
Digitization, 126, 148
Dillman, D.A., 117, 127, 202, 212
Direct marketing, 84
Direct-dialling, 15, 63, 72, 108, 145, 153, 155, 161; see also international
Dispersal, 33, 93
Dispersion, 15, 17, 35, 94–7, 187–90; see also regional processes, 34, 115
Distance, 3, 15, 33, 35, 36, 40, 64–6, 67, 71, 72, 83, 86, 88, 112, 124, 125, 161
  friction of, 13, 15, 157
  telecommunications, 98
  temporal, 34; see also global
Domestic, 39, 41, 48, 55, 56, 67, 68, 73, 84, 97, 98, 99, 100, 104, 105, 106, 110, 113, 143, 147, 150, 153, 155, 157, 163, 165, 181, 190
  calls, 21, 64, 65, 70, 153; see also local calls, barriers, cities
Dominican Republic, 183
Donau, A., 111, 210–1
Dordick, H.S., 142, 145, 202
Downs, A., 97, 202
Downtown, 54, 63, 108, 112, 113; see also CBDs
Drennan, M.P., 112, 202
Dublin, 103, 124
Ducatel, K., 38, 205
Dutch, 55, 64
Dymond, A., 157, 202

EARN, 184
EC, 101, 124–5
Economic, 4, 20, 58, 69, 83, 94, 101, 115, 128, 131, 136, 138, 139, 146, 155, 157, 160, 161, 168, 173, 174, 176, 177, 180, 186, 190, 191, 194, 195
  activities, xv, 7, 10, 14, 15, 33, 55, 94, 97, 108, 117, 128, 129, 155, 157, 181, 187, 188, 189, 194
  base, 101, 193

development, 5, 15, 73, 109, 117, 129, 130, 133, 134–9, 143, 145, 194
  growth, 106, 146, 147
  perspective, 33, 128, 180
  sectors, 7, 18, 116, 120–3
  structure, 124, 130
  ties, 66, 169, 173, 176, 178
  wealth, 135, 138; see also information, regional
Economies, 97, 100, 107, 112, 113, 196
  external, 139
  of scale, 53, 141; see also industrial, information, international, national, regional, urban, urbanization
The Economist, xviii, 22, 26, 27, 29–30, 63, 64–5, 66, 68, 79, 129, 142, 147, 164, 169, 183–4, 196, 202–3
Economy, 7, 97, 102, 105, 122, 123, 172, 173, 186, 191; see also global, market, rural, service, world, urban
Egypt, 177
Electronic, 14
  mail networks, 12, 57, 75, 180, 183–5
Elton, M., 85, 201
England, 124, 155
Equator, 40
Ergas, H., 165–7, 203
Erzberger, H.R., 123, 203
Europe, 24, 26, 40, 47, 48, 66, 79, 82, 93, 101, 102, 124–5, 134, 137, 146, 164, 169, 172, 177, 183, 184, 189
European, 24, 26, 29, 46, 65, 66, 72, 93, 101, 103, 124–5, 133, 137, 143, 165, 169, 172–4, 177, 180, 181, 184
EUTELSAT, 46
Exchange, 9, 143, 147, 160, 161, 192, 194
Exchanges, 4, 21, 25, 29, 35, 36, 53, 61, 88, 89, 99, 108, 153, 155, 160, 180; see also voice, global
Exports, 63, 99, 105, 178–9

Factories, 119, 121
Falk, T., 4, 14, 15, 33, 60, 108, 198, 203
Fax (facsimile), xv, 12, 20, 23, 26, 34, 57, 65, 68, 70, 73, 77–80, 84, 108, 141, 143, 157, 161, 162, 167, 168, 183–4
Federal Communications Commission (FCC), 38, 145, 165, 172, 173–5
Ferreira, J.H., 142, 200
Fibre-optic, xv, 4, 23, 25, 26, 36, 38, 47, 48–51, 52, 54, 55, 56, 72, 105, 110, 126, 190, 195, 196
Finland, 83, 175
Flexiplace, 83
Flexispace, 83

Flexitime, 84
Flow, 60, 139, 157, 161, 180
  barriers, 68–72
  characteristics, 64–8
  -induced patterns, 72–3; see also barriers, societies
Flows, xvi, xvii, 12, 14, 17, 60–73, 88, 101, 106, 128, 192, 194
  information, 14, 17, 32, 33, 60, 61, 63, 64, 66, 68–73, 88, 117, 122, 139, 161, 172, 194; see also social, societies, international, global
Fort Lauderdale, 115
Fox-Przeworski, J., 108, 203
France, xvi, 4, 25, 26, 76, 124, 133, 143, 147, 149, 155, 168, 169, 172, 179–80, 184
France Information, 22, 28, 39, 203
France Telecom, 147
Frankfurt, 103, 151
French, xvi, 4, 25, 27, 28, 38, 70, 115, 146, 151
Frenchboro, 126
Friedman, J., 99, 100, 203

G7 countries, 168, 169, 172, 175
Gappert, G., 99–100, 207
Gaspar, J., 178, 203
Geneva, 163, 184
Genosko, J., 121, 203
Geostationary circle, 40, 41
German, 70, 115, 146, 147, 151, 175
  -speaking, 70, 175
German Telekom, 147
Germany, xvi, 66, 70, 76, 121, 133, 143, 147, 149, 151, 167, 168, 169, 172–7, 179
Gershon, R.A., 46, 48, 203
Gillespie, A., 9, 12, 13, 14, 15, 24, 35–6, 40, 117, 120, 124–5, 203, 204
Global, 53, 66, 69, 96, 97–9, 102, 105, 109, 126, 140, 141, 144, 157, 168, 169, 172, 186, 190, 191, 192–5
  activities, 95, 96, 181
  capital, 100, 109, 151, 172, 179
  distribution of telephones, 130–4
  economy, 48, 52, 65, 66, 73, 97, 99, 102, 106, 114, 120, 144, 155
  finance, 101, 106, 109, 145
  level, 15, 190
  scale, 131, 144, 187
  services, 40, 109
  system, 97, 165
  traffic, 105, 167, 169
  urban hierarchy, 98–106, 187, 189, 193

urban system, 93, 97–110, 113, 155
village, 13; see also city
Glynn, S., 142, 146, 149, 150, 203
Goddard, J.B., 9, 12, 14, 24, 28, 35–6, 40, 85, 105, 108, 111, 114, 117, 120, 121, 124–5, 203–4
Gold, M.A., 201
Goldmark, P.C., 95, 204
Goldschmidt, D., 128, 204
Goods, 5, 7, 10, 61, 98, 99, 116, 119, 129, 160–2, 167; see also commodities
Gottmann, J., 6, 10, 14, 32, 33, 54, 95–7, 98, 108, 111, 204
Gould, P., 74, 204
Government, 5, 9, 10, 24, 38, 58, 61, 63, 72, 77, 99, 119, 123, 125, 138, 140, 141, 143, 145, 148, 153, 157, 180, 182, 196
Governmental, 26, 96, 119, 139, 141, 147, 162
  involvement, 9, 26
  monopolies, 149, 152
Gray, P., 211
Greece, 124
Grimes, S., 103, 204
Gross, D., 32, 204
GTE, 126
Gulf States, 177
Gupta, A., 81, 215

Hall, P., 129, 184, 204, 209
Hamilton, A., 102, 204
Hanneman, G.J., 55, 204, 211
Hansen, S., 119, 121, 124–5, 204–5
Hardy, A.P., 135–6, 205
Harkness, R.C., 85, 205
Hart, J.A., 24, 205
Heald, G.R., 210
Healy, P.E., 180–1, 214
Heng, T.M., 142, 205
Henry, W.A. III, 157, 205
Hepworth, M., xvi, 4, 9–12, 14, 35–8, 54, 55, 93, 102, 109, 115, 117, 121, 189, 196, 205, 213
Hierarchical, 37, 74, 184, 193
  pattern, 74–5, 80
Hierarchy, 29, 88, 97, 98, 99, 100, 189; see also urban, global, industrial
High-tech, 106, 121
  industries, 55, 147, 181
Hoffman, D., 157, 205
Hong Kong, 65, 100–1, 148, 169, 181
Hottes, K., 48, 205
Household, 80, 82, 138

members, 30, 58
sector, 129, 168
Households, xv, 17, 20, 32, 63, 80, 82, 89, 117, 125–6, 139, 141, 145, 153, 161, 162, 178, 185, 191, 193, 194, 195, 196; see also rural
Houston, 109
Howland, M., 112, 119, 122–3, 127, 182–3, 205
Hubs, 105, 146, 151, 165, 181
  global, 98, 100–6, 110
  regional, 55, 98, 100–1
Hudson, H.E., 128, 131, 142, 164, 205, 212

Image processing, 183
Imaging, 123
Imports, 63, 178, 179
India, 183
Indian Ocean, 40, 66
Indonesia, 46
Industrial, 55, 102, 98, 108, 121, 122, 182, 189, 192, 194
  economies, 9, 189
  park, 55, 115
  plants, 89, 97, 121
  production, 35, 93, 94, 101, 108, 147, 189, 192
  revolution, 5, 24; see also high-tech
Industries, 4, 5, 57, 124, 196
Industry, 4, 21, 54, 123, 139, 140, 141, 144, 146, 161, 166, 182, 183, 186, 187, 191, 195; see also manufacturing
Information, 3–19, 24, 25, 27, 30, 35, 40, 54, 55, 63, 64, 66, 72, 83, 84, 88, 96, 99, 100, 102, 105, 106, 109, 112, 114, 115, 120, 121, 123, 127, 129, 146, 157, 160, 164, 177, 180, 182, 186, 190–6
  commodification of, 10–12
  distribution, 93, 169
  economic, 61, 157
  economies/societies, xvi, 3, 6–12, 58, 89, 125, 126, 196
  exchange, xv, 153, 173
  institutional, 61, 157
  movement, 58, 161
  mover, 58, 108
  processing, 4, 9
  processors, 5
  produced, 61, 63, 157, 161
  production, 73, 114, 192
  services, 25, 99, 157, 194, 195, 196
  social, 61, 157

technologies, 4, 5, 6, 10, 13, 18, 54, 124–5
transmission, 4, 5, 35, 96, 98, 114, 120, 192
types, 61–3, 88–9, 154, 157, 161, 162; see also digital, flows
INMARSAT, 46
Innis, H.A., 15, 205–6
Innovation, 7, 21, 23, 58, 74, 75, 105, 127, 190, 196
Innovations, 21, 25, 75, 89, 140, 144, 189, 190, 191, 193
INTELSAT (International Telecommunications Satellite Organization), 40–1, 46, 47, 48, 66, 165
Interaction, 160, 161, 176, 192
Intercommunications, 4
Intercontinental, 48, 153
International, 41, 46, 54, 66, 73, 79, 98, 100, 103, 104, 108, 109, 110, 111, 113, 114, 143, 151, 155, 162, 163, 178, 181, 186, 187, 188, 196
  accounting, 153, 163, 165–7, 192
  agencies, 40, 51, 163
  arena, 12, 65
  calling, 68, 72, 164
  calls, 13, 63, 65, 67, 145, 165, 168, 172, 175, 179
  capital markets, 109, 114
  centres, 98, 102
  component, 99–100, 113, 144, 163
  dedicated networks, 180–5
  economies, 107, 108
  flows, 63, 66, 73
  gateways, 109, 165
  level, xvii, 20, 73, 83, 151, 193
  movements, 153, 160, 162, 177–80, 193
  networks, 55, 154
  telecommunications, xvii, 25, 39, 41, 47, 50, 53, 55, 61, 66, 69, 70, 72, 103, 105, 110, 142, 146, 153–86, 191
  telephone traffic, 105, 167
  telephony, 47, 155
  trade, 73, 102, 161, 162
  traffic, 166, 168, 169, 174, 191
  transmission, 21, 64–5, 88; see also capital, city, tourism, barriers, banking
International Encyclopedia of Communications, 3, 22, 206
Internet, 183
Interregional
  development, 73
  gap, 126

inequality, 118-20, 187
level, 85, 108
phone calls, 64, 67, 70
Intervening opportunity, 161
Iran, 173
Ireland, 124, 181, 182-3
ISDN, 25-7, 35, 54, 72, 195
Island Telephone, 126
Israel, xvii, 54, 63, 64, 65, 80, 83, 121, 148, 172-4, 178-9
Israeli, 66, 72, 148
Italy, 23, 26, 121, 124, 169, 172-3, 175, 179
Italians, 70
ITU (International Telecommunications Union), 41, 71-2, 131, 135, 163-4, 167

Jamaica, 183
Janelle, D.G., 15, 206
JANET, 184
Janssen, L., 64, 178-9, 212
Japan, 25, 29, 40, 47, 53, 65, 70, 105, 106, 110, 137, 145-6, 149, 166, 169, 172-3, 181, 184, 192
Japanese, 28, 103, 104, 106, 115, 146, 147, 183
Jensen-Butler, C., 178, 203
Jerusalem, 80
Jipp, A., 135, 206
Jowett, P., 4, 5, 206
Jussawalla, M., 128, 131, 180-1, 206

Keen, P.G.W., 151, 181, 206
Kellerman, A., 7-10, 17, 24, 29, 32, 56, 61, 65, 67, 70, 73, 75, 81, 83, 84, 94, 97, 112, 114, 144, 155, 157, 159-60, 162, 168, 169-72, 174, 176, 178-9, 184, 190, 206-7
Kern, S., 22, 32, 34, 155, 207
Kiesler, S., 183, 214
King, A.D., 103, 207
Klaassen, L.H., 70, 207
Klees, S.J., 210
Knight, R.V., 99-100, 207
Koppelman, F., 85, 207
Krakover, S., 112, 207
Kraut, R.E., 84, 207
Ku-band, 41
Kumar, A., 84, 112, 207
Kutay, A., 95, 98, 111, 112-3, 207
Kuwait, 133

Land-use, 193
conflicts, 54

Langdale, J.V., 38, 50-3, 70, 71, 75, 77, 99, 101, 103-4, 109-10, 115, 142, 145, 146, 150, 157, 180, 181-2, 207-8
LANs, 27
LaRose, R., 127, 208
Latin American, 173, 176
Lauder, G., 124-5, 208, 212
Leeds, 99
Leff, N.H., 129, 138-9, 208
Lehman-Wilzig, S., 95, 208
Leinbach, T.R., xvi, 130, 200
Lerner, N.C., 63, 208
Lesko, A.P., 14, 128, 142, 209
Less developed countries (LDCs), 128-9, 136, 139, 163, 168; see also developing nations
Less favored regions (LFRs), 124
Lesser, B., 129, 209
Lewis, L.T., 81, 209
Lewis, N.D., 157, 209
Liberalization, 143, 147, 151
Lines, 21, 26; see also telephone
Local, 73, 113, 114, 143, 145, 184, 187, 188
calls, 64, 66, 67, 71
level, 20, 108
London, 14, 38, 53, 55, 69, 97, 101-4, 114, 124, 145, 151, 165, 194
Long, M., 42-5, 209
Long-distance, 27, 33, 38, 39, 40, 51, 54, 56, 65, 71, 75, 106, 110, 127, 141, 143, 145
calls, 64, 127
market, 25, 77
Lord, J.D., 209
Los Angeles, 105, 109, 114
Low, L., 142, 205
Lyon, D., 155, 209

Ma'ariv, 27, 28, 29, 81, 82, 149, 209
MacMahon, A.M., 22, 209
Manchester, 93
Mandeville, T., 97
Manufacturing, 10, 15, 17, 34, 63, 94, 99, 108, 116-29, 140, 141, 188, 189, 193; see also industry
Maritime, see cables
Market, 27, 47, 50, 52, 68, 72, 84, 102-5, 138, 145, 152, 166, 182, 183
stock, 69, 104-6
Markets, 65, 104, 109-10, 113, 146, 167
international markets, 112-3; see also capital, global, regional, Japanese, urban

McCarroll, T., xv, 10, 12, 28, 76, 82, 209
MCI, 38, 50, 145
McLuhan, M., 13, 30, 209
Mandeville, T., 209
Martin, J., 65, 80, 209
Martin, M., 75, 119, 210
Marvin, C., 30, 77, 210
Mayo, J.K., 129, 210
Mediterranean, 48, 137
Megalopolis, 14
Melbourne, 150, 184
Mercury Communications, 38, 146, 150
Metropolitan, 95, 96, 114, 126
   areas, xvii, 56, 93, 94, 96, 97, 99, 111, 112, 114, 121, 125
   centres, 95, 96, 116, 123
   level, 15, 190; see also households
Mettler, J., 127, 208
Mexico, 65, 143, 169, 172, 184
   Mexican, 147
Meyrowitz, J., 30–4, 210
Mezzogiorno, 124
Miami, 109
Microwaves, 36, 38–47, 63
   antennas, 25, 53
   relay stations, 38, 53–4
   transmission, 24, 145
Middle East, 133–4, 137, 173, 177
Minc, A., 4, 147, 211
Minitel, 25, 27
Minneapolis, 114
Mintz, L., 210
Modems, 84; see also telephone
Montpellier, 184
Morgan, K., 189, 210
Morgan, M., 115, 210
Moss, M.L., 27, 38, 48, 54–6, 64, 74, 76–7, 83, 95, 100–6, 108–12, 114, 200, 210–1
Movement, 18, 32, 33, 60, 63, 83, 96, 113, 114, 154, 157, 160, 161, 177–8, 188, 189, 193
   aspects, 153, 157–62; see also information, international
Mowlana, H., 129, 211
Mukaida, L.V.D., 157, 209
Mullins, M., 166–7, 168, 171, 178, 201, 214
Multifunction polis (MFP), 115

National, 13, 26, 38, 66, 98, 99, 104, 106, 112, 122, 127, 130–152, 153, 157, 163, 165, 181, 183, 184, 186, 187, 188, 190, 193
   economies, 102, 181

   level, xvii, 83, 123, 14
   system, 38, 97
Natural monopoly, 24, 139–51
Netherlands, 66, 84, 169, 175, 178–9
NETNORTH, 184
Network, 24, 35–7, 53, 63, 72, 126, 146, 149, 150, 151, 196
   computer, 24, 36, 81
   structures, 35, 180; see also societies, technologies, global
Networks, xvi, 13, 17, 20, 25, 35–53, 60, 88, 96, 98, 112, 114, 115, 141, 146, 181–4, 190, 192, 195
   cable, 36–8
   dedicated, 154, 155, 180, 182, 191
   leased, 51, 53, 181
   public, 154, 180, 182, 183, 191
   telephone, 27, 36, 38; see also international
New Jersey, 55
New York, 38, 53, 55–6, 64, 69, 77, 97, 101–6, 113–5, 126, 151, 184, 194
Nicol, L.Y., 119, 211
Nicol, L., 97, 211
NICs (newly industrialized countries), 133, 189
Nijkamp, P., 68–9, 111, 211
Nilles, J.M., 83, 85, 211
Nippon Telegraph and Telephone (NTT), 146
Nodes, xvi, 13, 17, 20, 53–8, 60, 66, 88, 99, 102, 161, 190, 192
Nora, S., 4, 147, 211
North Africa, 134
North America, 26, 40, 46, 48, 66, 102, 103, 174, 189
   North American, xvi, 5, 14, 133, 136
Northern Ireland, 124
Norway, 175, 189
Noyelle, T., 113, 114, 211

Office, 12, 14, 53, 93, 111, 112,
   buildings, 54, 84
   suburban, 63, 193
   suburbanization, 111–3
Offices, 28, 34, 101, 188
   back, 15, 93, 95, 111, 113, 114, 116, 188, 193
   front, 113, 116
Offshore, 106, 127, 182, 188
   bank offices, 182–3
   data processing, 180, 182
Ogden, M.R., 128, 131, 206
Oklahoma City, 115
Olson, M.H., 84, 211

O'Neill, H.O., 105, 211
Ono, R., 163, 164, 211
Optical fibres, see fibre-optic
Optical scanning, 123, 183; see also fibre-optic
Orbit, 40
Organization, 5, 19, 26, 85, 130, 139, 140, 143, 144, 151, 163, 190, 196
  spatial, 17, 35, 189, 193, 195, 196
Organization for Economic Cooperation and Development (OECD), 5, 11, 180, 211
Organizational, 33, 46, 113, 130, 133, 151-2, 184, 186, 189
  paterns, 140-7
  systems, 38, 147
  transitions, 140, 148-51
Organizations, 37, 61, 112, 139, 162, 183

Pacific, 47-8, 50, 51-2, 65, 66, 157
  transpacific, 47, 48, 50
Pacific Rim, 48, 52, 58, 66, 102, 103, 106, 114, 133, 137, 169, 173, 189, 194
Pakistan, 177
Paris, 55, 100, 103, 124, 147, 151, 155, 194
Parker, E.B., 58, 117, 122, 126-8, 201, 212
Paterson, P., 165, 203
Peace, P., 113, 114, 211
PEACESAT, 157
People, xvii, 12, 17, 29, 33-4, 55, 58, 60, 61, 63, 70, 74, 83, 96, 99, 114, 153, 157, 160-2, 177-8, 189, 195
Penetration, xvii, 10, 29, 73-83, 124
Peripheral
  areas, 95, 188
  regions, 15, 34, 89, 116-29, 148, 188-9, 193
Peripheralization, 155
Peripheries, 17, 34, 89, 116-29, 131, 176, 189; see also rural, world
Perlmutter, H., 100, 212
Philadelphia, 77
Philippines, 177, 183
Phillips, A., 76, 142, 145, 212
Phillips, K.A., 55, 212
Phoenix, 115
Phone
  calls, 25, 28, 32, 68
  company, 26, 72
  companies, 27, 28, 66; see also telephone, interregional

Picturephone, 20, 23, 25, 28-9; see also videophone, videotelephony
Pitt, D., 189, 210
Pittsburgh, 105, 112
Place, 30, 32, 34, 84, 195
Places, 15, 29, 30, 33, 34, 60, 74, 75, 76, 85, 194, 195
Political, 69, 101, 104, 138, 139, 147, 155, 157, 160, 173, 174, 177, 180, 181, 196; see also barriers
Pool, I.d.S., 22, 25, 33, 41, 72, 73, 97, 139, 212
Portugal, 133, 176, 178
Post-industrial
  economy, 9, 189
  society, 6-7
Postal
  services, 3, 24, 146
  communications, 14
Pozo de Bisceglia, S.I., xvi, 212
POTS, 3, 27
Pred, A., 14, 75, 93, 212
Prestel, 25, 27, 76
Preston, H.F., 22, 202
Price, D.G., 80, 85, 212
Privatization, 145-6, 150, 193
Prodigy, 25
Product cycle, 182
Proussaloglou, K., 207
PTO (Public Telecommunications Operator), 146
PTTs, 24, 26, 52, 60, 140-1, 144, 147, 148, 152, 163, 165, 168, 180, 183
Pye, R., 28, 85, 111, 114, 124-5, 204, 212

Quarterman, J.S., 183-5, 212
Qvortrup, L., 57-8, 212

Radio, 21, 23, 61, 82, 157, 163
  stations, 165, 195
  -telephony, 163, 165-6
Ratzel, F., xv, 14, 212
Raulet, G., 32-4, 212
Razin, E., 117, 148, 213
RCA, 165
Real estate, 7, 53, 95, 111, 113-5, 118-9, 188
Region, 52, 100, 130, 177; see also metropolitan
Regional, xvii, 13, 14, 36, 38, 46, 66, 67, 99, 113, 114, 119, 120, 122, 127, 147, 169, 190
  development, 72, 116-29
  economies, 102, 113

gap, 127, 148
level, xvii, 15, 20, 83, 89, 108, 150, 193
scale, 54, 18; see also flows, nodes
Regional Bell Operating Companies (RBOCs), 70-1, 145, 196
Regions, 108, 120-9, 133, 150, 193; see also peripheral, world, rural, urban
Regulation, 138, 154, 162-7; see also deregulation
Residential, 53, 63, 74, 75, 136
areas, 61, 63
Retailing, 37, 82, 84, 114
Richardson, H.W., 128-9, 201
Rietveld, P., 64, 72, 176, 178-9, 211, 212-3
Robins, K., 12, 13, 14, 15, 203, 213
Robinson, A.L., 22, 213
Robinson, F., 117, 120, 124-5, 213
Robinson, P., 140, 149, 213
Rome, 184
Roscoe, A.D., 212
Rossera, F., 70, 72, 176, 212-3
Rothwell, M., 4, 5, 206
Rotterdam, 55
Rubinstein, A., 34, 213
Rural, 58, 117, 122, 127
areas, 57, 81, 127, 139
households, 58, 126
telephone system, 126-7

Salomon, I., 83, 84, 86, 111, 117, 148, 207, 211, 213
Salomon, M., 84, 213
Salt Lake City, 115
San Antonio, 55
San Francisco, 77, 109
Sarch, R., 163, 198
Sassen, S., 102, 104-6, 188, 213
Sassen-Koob, S., 100, 213
Satellites, 4, 13, 20, 21, 23, 25, 36, 38-47, 54, 63, 66, 81, 129, 155, 163, 165, 178, 185, 190
Saudi Arabia, 173, 177
Saunders, R.J., 117, 128, 135, 138-9, 142, 147, 213
Sawhney, H., 126, 213
Scandinavian, 57, 58, 133, 136, 175
Scanning, see optimal scanning
Schieber, L., 12, 200
Schmidt, S.K., 142, 147, 214
Schneider, H.N., 213
Schofer, J., 213
Schwartz, A., 114, 214
Scott, A.J., 182, 214
Scottland, 183

Semiconductor, 23
Seoul, 184
Service, 9, 47, 55, 56, 74, 82, 114, 115, 116, 122, 123, 126, 127, 131, 133, 138-43, 148, 149, 165, 166, 178, 189, 193
activities, 93, 94, 111
economies, 6-9, 18, 54, 56, 58, 99, 144, 147, 190
functions, 115, 123, 182
industries, 140, 188
providers, 25, 120, 141, 196; see also governmental, international, radio, telephone
Services, 3, 4, 5, 6-12, 14, 30, 35, 40, 46, 54-6, 61, 70, 74, 77, 82, 88, 94, 100, 101, 104, 106, 109-12, 114, 116-29, 131, 133, 135, 138-52, 155, 157, 167, 185, 187-8, 191, 193, 195-6
accounting, 7, 55
advanced, 54, 56, 57-8
business, 8, 94
financial, 55, 69, 100, 181
legal, 7, 55, 114
producer, 54-5, 94, 97, 100-1, 108, 111-5, 120, 122, 141, 188-9; see also global, information, postal, regional, transportation, computer
Sheffer, D., 121, 214
Shopping, 83, 84, 112
Short, J., 29, 95, 214
Siemens, 132, 134, 137, 153, 214
Simpson, A., 22, 77, 214
Singapore, 65, 100-1, 181
Singlemann, J., 131, 214
Sioux Falls, 115
Smart buildings, 20, 54-5, 88, 99, 100
Smith, K.A., 180-1, 214
Snow, M.S., xvi, 4, 5, 41, 46, 214
Social, 5, 17, 20, 29, 30, 32, 58, 61, 68, 74, 128-9, 131, 138, 139, 146, 155, 157, 160, 162, 174, 179, 180, 181, 186, 190
aspects, 17, 30
calls, 66, 129, 178
contacts, 32, 157, 161
effects, 13, 30
life, 18, 68
relations, 63, 195
sphere, xv, 34
ties, 32, 33, 61, 66, 89, 173, 178; see also barriers, place, services, information
Socializing, 83, 84
Socially, 33, 70, 173, 191, 195

228 Index

Societal, 6
societies, 14, 32, 121, 196; see also information
Society, 5, 7, 17, 60, 138, 189, 194–6; see also flow, post-industrial
Sociospatial, 89
characteristics, 21, 30–5
Sonderegger, U., 123, 203
South America, 48, 134
South American, 137
South Korea, 183, 184
Soviet Union, xv, 155, 164, 176, 177
Soviet, 41, 157, 177
Spain, 176
Sproull, L., 183, 214
Sri Lanka, 58
Stanley, K.B., 166, 214
Staple, G.C., 14, 47–8, 52, 66, 103, 142, 147, 153, 167, 168, 169, 171, 178, 214
STAR (Special Telecommunication Action for Regional Development, 124
Suburban, 111, 114
areas, 113, 188
location, 113, 188; see also office
Suburbanize, 34, 95, 113
Suburbia, 15
Supply, 7, 56, 64, 96, 125, 139–1, 148, 149, 162, 166, 178, 194
Sweden, 14, 26, 29, 58, 83, 131, 169, 174–6, 181
Swedish, 14, 133, 181
SWIFT, 109, 182
Switches, 4, 23–5, 126; see also telephone
Switching, 23–4, 26, 35–7, 72, 88; see also exchanges
Switzerland, 66, 70, 101, 133, 163, 175, 179
Swiss, 101, 123
Sydney, 150

Taiwan, 65
Tampa, 115
Taylor, L.D., 30, 214
Technological, 4, 5, 7, 20, 23, 47, 54, 58, 69, 77, 89, 95, 104, 106–9, 112, 119, 123, 140, 141, 144–5, 161, 166, 178, 183, 186, 188, 190, 193; see also barriers
Technologies, 3–5, 15, 17, 20, 21, 24, 36, 65, 74–5, 86, 88–9, 94, 95–6, 105, 113, 115, 127, 129, 131, 141, 154, 168, 178, 180, 189–90, 196
network, 21, 36; see also information, switching, transmission

Technology, 7, 27, 29–30, 38, 41, 47, 48, 59, 68, 72, 74, 79, 80, 82, 86, 93, 124, 126, 129, 144, 160–1, 163, 183–4, 190, 195–6; see also digital
Technopolis, 115
Tel-Aviv, 80, 148
Telecom, 150
Telecommunications Development Bureau (TDB), 164
Telecommuting, 83, 85, 94, 123
Telecomputing, 73
Teleconferencing, 28, 86, 94; see also confravision, videoconferencing
Telecottages, 20, 54, 57–8, 88, 99
Teledensities, 131–8, 148, 151, 164
Telegeography, 14
Telegrams, 108
Telegraph, xv, 21, 23–4, 77, 108, 144, 155, 167, 190
Telematics, 4, 13, 93, 121
Telemedicine, 129
Telemex, 147
Telephone, xv, 4, 12, 14, 20, 21, 23–5, 27, 29–30, 32–4, 40, 47, 66, 68, 70, 73–7, 84–86, 107–8, 126, 136, 144–6, 155, 160–2, 165–6, 168, 179, 190–1, 196
calls, 13, 60, 63, 85, 128, 191
companies, 14, 24, 27, 38, 60, 65, 71, 126–7, 153, 168, 183, 195, 196
conversation, 29, 34
density, 14, 133
lines, 4, 25, 26, 29, 54, 63, 75, 131, 157, 169, 196
service, 24, 30, 68, 76, 77, 125–8, 131, 139, 140, 141, 186
system, 25, 37, 56, 63, 70, 79, 127, 130, 144, 155
systems, 26, 27, 126, 133, 151, 168; see also long-distance, networks, phone, residential, rural, digital, switches, modems, exchanges, international
Telephones, 4, 14, 26, 28, 32, 130–1, 136, 151, 162, 168
Telephonies, 24
Telephony, 51, 126, 135, 167, 195
voice, 25, 143, 180; see also radio, international
Teleports, 20, 54, 55–6, 57, 88, 99, 100, 105, 187, 190
Teleshopping, 84–5, 94
Teletext, 82
Television, 4, 20, 23, 28, 36, 40, 51, 61, 75–6, 81–2, 99, 190, 192, 195–6
Television Factbook, 82, 215

Telex, 12, 21, 23–4, 79–80, 108, 124, 155, 160, 167
Temporal, 15, 30, 32, 34, 66, 69, 108, 195; see also barriers
Temporality, xvii; see also barriers; distance
Thailand, 65
Third World, 51, 133
Three-legged stool, 97, 102, 105–6, 172
Thrift, N., 104, 215
Time, 15, 30, 32, 34, 57, 60, 64, 66–8, 74, 83, 85, 86, 88, 112, 151, 177, 195
differences, 160, 161, 169
-space convergence, 15, 32, 33
Toffler, A., xv, , 5, 13, 83, 131, 145, 215
Tokyo, 69, 97, 101–3, 106, 145, 151, 184, 194
Toong, H.m.D., 81, 215
Topology, 35–6, 53
Tornqvist, G., 14, 215
Tourism, 61, 99–101, 114, 125, 157, 172, 178–9
Towns, 94, 108, 193
Trade, 7, 17, 104, 109, 151, 172; see also international
Traffic, 40, 66, 70, 83, 88, 102, 128, 148–9, 154, 155, 165, 167, 169, 176–7, 187, 189; see also international, global
Public, 167–80
Transaction, 10, 98, 106
Transborder data flows (TBDF), 69, 180–1
Transferability, 161
Transistor, 23
Transmission, xv, 3, 4, 9, 21, 24–7, 29, 35–6, 38, 41, 47, 50–1, 58, 61, 63, 77, 79, 82, 108–9, 121, 129, 161, 162, 165, 168, 178, 182, 184, 189
cableless, 13, 36
data, 4, 20, 25, 27, 40, 47, 58, 68, 72, 108, 141, 143, 146, 155, 162
media, xvi, 12–3, 17, 20–35, 60, 63, 88, 190; see also information, microwave, networks, data, international
Transnational corporations (TNCs), 38, 53, 97, 100–1, 103, 151, 180–3, 194
Transportation, xvii, 7, 9, 12, 15, 55, 58, 83–6, 112, 116, 117, 119, 139, 141, 161–2, 182, 189
networks, 38, 97
Travel, 15, 17, 32–4, 55, 83, 86, 89, 112, 160
air, 55, 86

business, 28, 85
Tulsa, 115

UK, xvi, 23, 25, 26, 29, 40, 53, 66, 76, 80, 82, 99, 104–5, 110, 114, 124, 133, 143, 145–6, 147, 149, 150, 166, 169, 172–3, 175, 189, 193
US, xvii, 10, 12, 14, 21, 23–5, 27, 29, 36, 38, 40–1, 50, 52, 53, 56, 63, 64–5, 67–8, 70, 75–6, 80–2, 84, 97, 99, 103–6, 109–10, 112, 114, 125–6, 131, 136, 140, 144–5, 149–50, 157, 164, 165–9, 172–4, 176, 177–80, 181–2, 184, 189, 190–2, 193
US Bureau of the Census, 76, 81, 82, 84, 103, 215
US Federal Communications Commission (FCC), 22, 82, 215
US Sprint, 50, 52–3
Ullman, E.L., 161, 215
United Nations, 12, 104, 131, 163
Urban, 14, 38, 53, 61, 66, 93, 94, 108, 117, 122, 127–8
areas, 81, 94, 129, 138, 193
centres, 35, 96, 108, 115, 119, 121, 122, 125, 127, 128
economies, 93, 99–100, 107, 113
hierarchy, 61, 74, 77, 88, 94, 99, 102, 115, 189
level, xvii, 83, 192
life, 32, 94
scale, 54, 187
systems, 93, 96, 111; see also global, America
Urbanization economies, 112, 122
Ure, J., 142, 148, 215
Use, 30, 33–4, 88–9, 154, 162, 190, 192
Utilities, 7, 24, 139–41, 191, 196
Utility-penetration paradox, 74–5

van der Weg, A., 207
van Nierop, J., 212–3
VANS, 25, 27
VCRs, 75, 82
Venezuela, 179
Video, 25, 36, 39, 47, 63, 86, 108, 109, 131, 143
Videocamera, 28
Videoconferencing, 20, 25, 27–8, 54, 70, 86; see also confravision, teleconferencing
Videophone, 28–9, 70; see also picturephone, videotelephony
Videotelephony, 27–9; see also picturephone, videophone

Videotex, 27, 76, 84
Villages, 131
Voice, 3, 24–5, 27, 29, 36, 38, 47, 48, 51, 68, 70, 131, 162; see also telephony

Wadsworth, S., 204–5
Wagenaar, S., 207
Wales, 124
WARCs (World Administrative Radio Conferences), 41
Warf, B., 56, 105–6, 109, 114–5, 116, 215
Warford, J.J., 213
Washington, D.C., 38, 77
The Washington Post, 156, 215
Webber, M.M., 95–6, 215
Wellenius, B., 213
Williams, E., 214

Williams, H., 120, 203
Wilson, A., 46, 215
Wilson, L.J., 129, 211
Wilson, M.I., 182–3, 215
Wise, A., 97, 215
Wolff, G., 99, 203
World, xvi, 5, 15, 104–5, 131, 133, 151, 164, 168, 193
 cores, 15, 48, 66, 106, 169–74, 176, 189, 194
 regions, 48, 164, 173, 176
World Space Industry Survey, 46–8, 51–2, 66, 215
The World's Telephones, 153, 169–71, 215

Zurich, 123